A Conservationist Perspective

A Conservationist Perspective

By

Alan Weatherley

Library of Congress Cataloguing-in-Publication Data

Weatherley, Alan.
A Conservationist Perspective
ISBN 978-1-257-64428-5

To my wife Robena with love and thanks

CONTENTS

PREFACE

This book's intention is to encourage people's interest in the conservation of Nature, helping them to become more skilfully engaged as serious, long-term conservationists, and in awakening what I believe is a very real, but often dormant, love of the natural world. One of the central points I have tried to make is that if the conservation movement can actively engage the attention, long-term interest and informed labours of really large numbers of the public, this will result in much greater prospects of saving what remains of the natural world from ultimate degradation. I will add that, for conservationists the world around, the need for purposeful action is great and the time is very urgent. It is in the light of these facts that I am anxious to make what contribution I can.

Alan Weatherley
Professor Emeritus
University of Toronto
Autumn 2011

INTRODUCTION

APPROACHES TO CONSERVATION

"Conservation" was once a term commonly used in referring to unique or antique buildings, manuscripts or works of art. Now it very often refers to the entities and creatures of the world of Nature. And the people who are seriously interested in Nature conservation are universally referred to as "conservationists." There are now thousands, perhaps millions, who thus refer to themselves, but there is growing concern among them that they will, in fact, be unable to "conserve" Nature, because so many species have already become endangered or destroyed, at rates that are in many cases constantly increasing. Major problems include getting people to look at Nature with fresh eyes and really appreciating the magnitude of the damage, and then puzzling about just how to prevent further damage or, in some cases, effect repairs.

Contemporary conservationists, while having love and admiration for Nature, must guard against soft and sentimental views, and remedies for problems that have been too casually applied. Instead, we stress that humans have been a part of Nature, and indeed that for a very long time Nature was, and really remains, our one true home. It is, of course, a huge home with many rooms that most of us will never visit. Yet what we know of Nature's mysteries and grandeur tells us that its rooms are connected in such a way that bad housekeeping of any of them may affect other rooms in the overall structure, sometimes rooms close at hand, sometimes at great distances. In calling ourselves conservationists we are announcing our wish to safeguard as much as possible the rich variety and abundance of the natural world. We will use all our strength and ingenuity to avoid losing it. For to lose Nature, we believe, would be a profound denial of our own human selves as natural beings.

Conservationists would prefer to avoid major confrontations and conflict with those whose interests seem opposed to ours. But they also recognize they can succeed at times only if they steadfastly confront the more egregiously harmful of those industries, businesses and political measures and people that display an imperious determination to exploit the resources of the natural world to their absolute limits, while also failing to safeguard wildlife, what remains of fisheries, forests and productive farmlands, ignoring pollution of marine and inland waters and of the atmosphere, and declining to safeguard the wonders of the natural physical environment.

Dedicated conservationists call for the establishment of very strong standards of environmental quality to be adhered to by all citizens. In doing this, they must ensure their judgement and the quality of their advice and stewardship can be relied on. Further, the conservationist view of Nature is worth nothing if it is not that of idealists. Despite our own age-old tenancy in Nature we have, until recent decades, understood it but little in any except empirical ways. Great errors in how we have managed the planet have resulted. And failures will continue until we realize there is more love and regard for Nature than is usually believed. In this book I want to suggest how conservationists' main concern should be to get these feelings front and centre in the minds and acts of the general public. In fact, we must exercise the knowledge, will and determination not only by finding the necessary financial and administrative support, but especially through attitude and spirit. Nature's beauty has been the province of poets, artists and aesthetes for a long time. But their views of Nature have often been inflated, grandiose, overly heroic—the Romantic imagination in hyperdrive!

Nature can be grand, but is often reserved, steely and subtle in its grandeur. Its magnificence can also be overwhelming rather than decorative, pretty or cozy. It is not the ambition of true conservationists to replace Nature with artificially cultivated flower gardens and domestic pets! Their task and their responsibilities should be viewed with an analytical directness as well as with devotion.

Conservationists have a deep and wide affinity with the great natural features of the world (forests, oceans and mountains are some of the special examples), yet somehow our species has spent much of its history degrading or destroying these wonders. Recently, the eminent biologist Edward O. Wilson estimated that by the end of the

21st century about 50 per cent of present living animal species may be gone. Forever! A sad note. How is this tragedy to be explained and how should conservationists find ways to convince people that to damage Nature is going against our own deep moral, emotional and ethical interests? We need to face these questions immediately.

This book discusses how and why our relationship with Nature manages to accommodate itself to excessive destruction, and examines approaches that conservationists can adopt in attempting to rescue the present world. However, even though people may often know through everyday observation or experience that the creatures of Nature are threatened, not only by industry but by too much hunting, fishing, bad land use or forestry practice, by toxic agents, and by broad environmental damage such as global warming, the severity of impacts must be demonstrated by real evidence.

Indeed, if conservationists are serious about preserving wildernesses—including thousands of wildlife species, or in some cases just a few that are uniquely threatened—they must, no matter how well intentioned, turn to the science of ecology in order to be taken seriously. Ecology should dominate the toolkit conservationists need in order to reveal or authenticate damage or threat of damage to Nature.

Ecology, like physics, chemistry or mathematics is, as a science, largely value-free. Its main function is to determine how the life forms of the planet function in their wild settings. Conservation itself is full of value judgements and concerns for ethical and aesthetic considerations. So conservationists are not just ecologists with another name (though many ecologists are ardent conservationists), but serious conservationists will always call on ecological insights in order to preserve from damage or extinction whatever species, ecosystems or communities they see as threatened.

Sixty years ago, many ecologists were satisfied to help the advance of their science by investigating anything that threw new light on life cycles of plant or animal species in relation to environmental conditions. But during the past 40 years, the technical aspects of ecological research have advanced enormously. A myriad of facts that had seemed quite unavailable has now come within reach. My aim is the modest one of showing at a broad level the ways in which ecologists think about and approach their problems. To facilitate this task I have described case studies of which I have had direct experience. First comes a sketch of the ways in which childhood interests in live animals and an "outdoors" world led to ecological interests and outlooks. This is followed by a first opportunity to view the investigative methods of science as conducted by a very able scientist (who happened to be not an ecologist but a human physiologist). These experiences may convey something of how the air of discovery arises and grows, and encourage others to think about conservation questions in a scientific way, even though this is not all there is to conservation. Most of the studies described centre around ecological investigations of the state of originally natural systems that have been impacted by human activities. Understanding what was done to them seemed to me a way of bringing home the essential ecological nature of the systems that were considered, and possibly figuring out ways to conserve them from further damage. I hope readers will agree.

A further topic examined, also related to conservation, is human societal sustainability (and in particular self-sustainability). In the future many factors such as decreases in global space, food and natural resources will make it vital to plan far more carefully than is presently popular. Also discussed under this head is the vexed question of the meaning of societal wealth and how it is that some societies have become monetarily wealthy and militarily powerful, while others stay weak and poor, and are often exploited. Other matters discussed briefly in this book touch on questions of land ownership and accompanying responsibilities, the contrast between good and bad citizenship and the ethics of conservation.

I have tried to keep my approach to conservation and its accompanying ecology modest and readily understandable. The majority of people who are becoming concerned about conservation are involved in local matters: a lake threatened by pollution, clear-cutting of a woodland community, a coastal landscape undergoing irreparable damage from misuse. If a conservation challenge is huge enough, such as safeguarding a national park, a fishery or a large nesting colony of birds with complex breeding behaviour, then government conservation authorities may become engaged. And they will have the trained personnel, the numbers, the scientific equipment and the money, plus the force of a government to see things through. But smaller projects are also of vital interest. If they are not attended to the countryside of Nature becomes dissected into a ragged chequerboard of damage. However, if people can raise enough money to work effectively on a project they may be able to engage a competent environmental consultant to design and direct the work needed. But even then, the conservators should know enough to give broad direction to the professionals. Otherwise their essential aims may become misdirected and move in a different direction than what they want. If amateurs are resourceful enough they will train themselves in the ways of ecological analysis of conservation problems and become better able to see what is needed and how to get it done. Helping people to understand how to do this is the aim of this book.

Conservation of the world of Nature involves human welfare, clean water, a right to breathe pure air and enjoy diverse countrysides and landscapes with natural vegetation and a rich variety of wild animals. But Nature is less mighty than it was only a century ago. Once, humans were a direct part of Nature. Now we seem, in many countries, cut off from it, its power over us slight and distant. And even now, millions still sense that it has a power to touch them, though most would not think of themselves as conservationists. Yet somehow those who are devoted conservationists are stuck in a rut. They know that despite the popularity of their cause (to judge from opinion polls) they are managing far too rarely and slowly to get the results they want. And this is mainly because broad concern gets sidetracked whenever business, industry and political considerations insist that serious conservation will cost too much in taxes and in job losses.

Of course, much conservation, like attempting to restore ruined fisheries and preventing the further loss of the world's freshwater supplies, topsoil, forests and spread of deserts, will be very costly. But so what? Nature, with its remaining plant and animal life, can only be saved from eventual disappearance if humans collectively resolve to become conservationists. Surely the prospect of past and present damage being allowed to continue is terrible enough to tell us that we have to pay if we are to avert total tragedy. Unfortunately, this is not something industry, business, and politicians want to hear or face. The essential ambition is to make a profit. To maintain profits business can coldly contemplate the fate of the planet and will consistently assert the conservationists' case is hugely exaggerated.

The well of information is essentially bottomless. Think of an ecological problem and, providing you can pay for it, you can locate experts who can find answers for you.

But conservationists will not succeed if they do not address the real threats to the natural environment. The search for oil and minerals never ceases and both activities are stripping the earth of its soil and its rocks, even blasting the tops off mountains to expose coal. In less than another century the world we have known will be mostly a thing of memories. Only the shreds and death throes of the world we now live in will remain to tell all those with any sensibilities what wonders have been denied to them.

Conservation includes not only wild creatures, forests and all kinds of vegetation, birds, mammals, reptiles and fish, together with countless invertebrates, but also the freshwater that is being wasted by industry, by ever-expanding populations and by appallingly aggressive agricultural practices.

What are conservationists, really? What is the responsibility they wish to assume? And for how long and with what determination will they struggle to assume it? They must be the vanguard of those who move away from casual, soft and sentimental views of Nature to something with a harder, more absolute purpose. But they will not succeed unless they can bring the public to see that Nature's destruction is not only due to ignorance and neglect, but comes mainly from deep, sinfully offensive acts that result from hubris, greed, ignorance and fear. Their duty is to win a war of ideas and that will call for unshakable determination.

CHAPTER 1

RAW BEGINNINGS: A PERSONAL JOURNEY OF DISCOVERY

I was born in Sydney, Australia in 1928, not far from where Captain Arthur Phillip and the First Fleet landed in Camp Cove, 140 years earlier in 1788. The Sydney Harbour Bridge, world-renowned for its majestic steel arch, was in the first stages of construction, still just a dream on the Sydney skyline.

The city is gloriously sited, around its magnificent harbour, open to the sky and the ocean. I remember it as a place that could be subjected to torrential rains and powerful winds, sometimes savage hailstorms, but mostly the sun seemed to shine. It shone hot and white, with a ruthless summer power to burn those as fair-skinned as myself.

I loved the outdoors, including the great beaches. And those great outdoors, above all, are what I still remember and feel about Sydney. Though I've since lived in various places around the world, none compare to Sydney, with its light, its blue skies and ocean.

In that early part of the 20th century Sydney had a population of a million and a half people. It was Australia's chief metropolis. Its sprawling domestic scene was of thousands of brick bungalows with roofs of red or brown terracotta tiles, their lawns and gardens defiant against the drought-prone climate, surviving on inhospitable soils derived from that same sandstone that defines the whole city and urban environment.

There was a crowded mercantile core, but most of the surrounding suburbs that sprang up after WWI were open, lacking any of the closeness and settlement of most large cities.

Kids roamed and biked the hills, played in the parks and on the beaches. There were eucalyptus trees and wildlife—lots of birds, lizards (skinks), Christmas beetles and cicadas that sang resonating songs, which rang in your head, semi-wild cats that survived in the mild winters and raided garbage cans and drove us crazy on summer nights with their snarling mating fights. Ants were everywhere.

Looking back at it after so many years it seems close to having been that dream city that did not displace the ancient soil, rocks, native trees, or wildlife, only encompassed them, contained them, mixed itself up with them, shared their life and landscape. The expanding edge of Sydney's suburbs was as near to my vision of urban environmental harmony as I would ever know.

I must have been three or four when I first began to dig in the garden for earthworms that I would stretch to their limits like rubber bands, until they broke and their sticky, wet insides spurted on to my fingers. I rolled over bricks, rocks and pieces of wood looking for slaters. Sometimes I dislodged large centipedes or spiders.

Every December came the Christmas beetles, large and beautiful bronze-gold scarabs of several species, strong of limb, great fliers, thronging the street lights at night.

I became seriously ill when I was five and took a long time to recover. To re-engage my interest in living creatures and things outside myself, my grandfather manufactured an aviary for canaries from a large wooden packing case. He equipped it with perches, feeding stations and a galvanized steel floor sheet that could be removed and hosed clean. It was mainly my task to take care of half a dozen canaries, feeding them on bird grain, lettuce, and the dandelions that I had picked for them, keeping them watered, their cage clean and supplied with cuttlefish bone and grit. I embraced my task and I would watch the birds for hours as they flitted around, singing. Then two of the canaries announced their sex by laying eggs in the small cans that had been installed for them to make their nests in. I was fascinated and anxious and kept daily watch. The eggs hatched and canary chicks entered the world. Unfortunately, though apparently healthy at first, the chicks died off one by one. With my parents' assistance I tried to nourish them with tiny morsels of hard-boiled egg, but to no avail. The smallest and most ill formed of the chicks, which my father with misplaced good humour named "the runt", was the last to die. I was much distressed to watch its final struggles.

By the time it was all over I had learned things about life and death, seen how small birds behaved and gained a sense of the meaning of life cycles. I didn't give serious names to these things, but they were lessons that stuck.

At seven, I found a tiny, naked, blind mammal, perhaps an inch long. Ants were attacking it, but still it moved feebly. I took it to my grandmother, a countrywoman I loved and whose fund of knowledge I had not grown old enough to doubt. Like most rural people of her period she really knew little of marsupials, had never seen a naked juvenile one before and, no more than I, did she recognize what she was looking at. Just down the road from our place was a palatial house in whose dense, protecting camphor laurel trees lived a colony of ringtail possums. They were almost certainly the source of this pathetic, dying mite, as I later realized.

Sydney Harbour sparkled in the almost endless sun, and around its Eastern suburban shores were thousands of tidal rock pools in the sandstone, occupied by anemones, small fish, shrimps, mussels, sponges, crabs, starfish, gastropod and bivalve molluscs and seaweeds of many forms. At low tide, I would tramp from rock pool to rock pool, gulls circling above, and sometimes I heard the lions roar across Sydney Harbour in Taronga Park Zoo. It is not too fatuous, I hope, to say that I was not a learned child but yet was gripped by "the glories of Nature" that were so evident to anyone with eyes to see.

Ants were the first animals I began to pay attention to. I used a magnifying glass to concentrate the powerful sunlight into a death ray. I would pick on individual ants and focus the heat on them as they rushed along trails in the grass or over bare ground. They could escape down a hole or in deep grass or into shade, but otherwise the death ray would get them and they would die squirming. I would roast the corpses until they sizzled and smoked.

After a while I got tired of being an ant hunter and turned from small black ants to the much larger red-brown "mound" ants that made their nests in grass, in gardens, beside rocks, wherever they could get covering materials to protect their mounds. It was easy to rouse them enough to come boiling out of their nest openings. They would attack my feet and bare legs, but had no sting, though my bare legs and feet might smart a bit from the myriad tiny wounds inflicted by their mandibles. Soon I was dropping grasshoppers or beetles on the excited ants to see what would happen, but neither grasshoppers nor beetles could be stopped from moving away without harm. To really see what the ants could do to other animals that couldn't escape I put both ants and prey in glass jars to watch for hours as they eventually overwhelmed the prey, held them down by all their legs, and began to disable and kill them.

At last these gladiatorial encounters became boring as they took on a feeling of predictable, almost formal, demonstrations, and I looked for things more like genuine battles. I had never seen ants of one species attack those of another, but I knew of one species in which the individuals were of a size similar to those I had been "testing." I collected samples of both—the "dark reds" I had been watching and also of a "ginger-red" species of slightly larger individuals—and put them in glass jars together, varying the ratios of their numbers to determine which species was the stronger in battle. Whatever the childish savagery of the impulse to stage these struggles, I spent hours watching their outcomes and did establish that the slightly smaller "red-browns" were stronger, had a more lethal bite, and were superior in battle to "ginger-reds." I would later realize that this was my first conclusion that had been based on something that crudely resembled a simple, but science-like, experiment.

When I was twelve I got another lesson in biology. In a 19th century book titled *Natural History for Young People* that I discovered in a cupboard at my grandparents' house, I found instructions for keeping tadpoles and the frogs they would become. I caught some large tadpoles in a pond in a nearby park and installed them in a large glass jar with a series of platforms on which the frogs could dwell after they were formed. The food I offered the tadpoles was small breadcrumbs. Not, I think, an adequately nutritious diet, for after some weeks, though they began to sprout legs, lose their tails and gradually approach a true froggish form, all but one died. The sole survivor did at last become a complete-looking baby frog, and actually dragged itself onto the lowest of the platforms. There it rapidly expired. This was, again, something of the same sort of lesson I had from the canaries. In both cases the old adage applied: "The operation was successful but the patient died." But at least I had witnessed vertebrate animals going through important parts of their life cycles.

What else was there in the "biology" of those days? Visits to the zoo, with the colossal fun of watching primates—especially a chimpanzee hanging upside down from its cage roof and urinating slowly and deliberately into its own mouth while watching several kids howl with glee—and the chimp with its own apparent sly appreciation of its feat. And another chimp hurling a lump of its faeces, quite accurately, at a watching nun. Poor nun. Merciless, open laughter from some kids, muffled sniggers from others (were they Catholics?). Monkeys and apes were more than just "clever and amusing" and very distant relatives of ours. They appeared quite aware of their actions and apparently appreciated their consequences (even if it was embarrassing for nuns).

Otherwise there was the sheer magnificence of the range and variety of shape, size, colour and behaviour of mammals, birds and of marine creatures in the splendid aquarium.

And what else? Maybe during Australia's 150th Anniversary in 1938 when my uncle caught a shark outside Sydney Harbour in an ocean fishing contest that was part of the Australian sesquicentennial celebrations. The shark, a hammerhead of record size, hung on display from a wooden scaffold at Watson's Bay, only a couple of miles from where I lived. It looked monstrous, still menacing. A glimpse of alien power and beauty in death as I looked up at it against the brilliant sky.

Perhaps I began to be a biologist on that bright afternoon as creatures everywhere, people as well as animals, seemed to swim and sail and rise and sing in the blue glare of summer under that white and searing sun.

Though I grew up in a big city, I saw more animals at close range than many country kids. And unlike them, I was free of the need to worry about whether the animals were useful and beneficial, or pests, or dangerous, or in competition with me or my family for our livelihood. I could remain objective, my activities towards them fuelled only by curiosity and interest. I was able to see animals and later plants, in terms of their intrinsic biological characteristics.

I grew up knowing no professional biologists. For most male kids of my time, studying science at school meant five years of chemistry and physics, culminating in matriculation. I will not comment on my school education in science except to say that it was driven by far too much parroting of rules and application of standard formulas in the solution of routine "problems" requiring little or no insight or real understanding, and too little observing and learning about the manifold phenomena of the natural world, of pondering what things are and how they work.

Someone like me, with unclear attitudes about biology as a serious subject (mostly taught at girls' schools as a scientific "soft option"), but interested in it, would need to study it in a form perceived to have some "seriousness," that is, related to the applied science fields of agriculture, medicine, veterinary science, dentistry or pharmacy. That would validate its integrity, legitimacy and significance by being associated with a useful profession. And in the Australia of the immediate post-WWII years, agriculture appeared more than ever to be a major part of the country's future, its national prestige, its mystique as a society, its reputation. It was in this realm of national activities that the "life sciences" were seen as appropriate and necessary tools of great national endeavours. The society of Australia would, of course, soon undergo substantial changes of sorts not easily perceived at the time.

I began to attend Sydney University as a naive sixteen-year-old and, with the idea of getting a broad exposure to biology, got enrolled in the Faculty of Agriculture.

I quit Agriculture at the end of the year. It had been a false start for me. Put it down to utter immaturity. However, even this brief stint in Agriculture opened my eyes to things I had never before thought of. Our first year required us to spend a month on farms for practical experience. I put in three weeks on a dairy at Camden, about thirty miles out of Sydney and one of the earliest places to be farmed in New South Wales.

Even for one of my innocence of country ways the ravages to soil in the form of erosion from aeons of overgrazing by sheep, and sometimes cattle, were glaringly obvious. Nor did it have to be explained that the original mature forest had been stripped from the land long before. And finally, populations of rabbits were there in their pestilent abundance.

But there were other disturbing things to note. I had taken for granted that all farmers, even the not very good ones, would take care of their animals and try to be good stewards of the land. Yet the very farmer whose land I was working on averred forcefully that "all trees should be cut down," indeed that "trees caused erosion." This same man, in a fit of total fury caused by a laggard cow that declined to go smoothly to her milking station, brought the animal to her knees with a ferocious blow between the eyes from a heavy broom. I had not become a great admirer of domestic cattle; indeed I shared the farmer's frustration with this reluctant animal. But his attack was stupid, brutal and hardly calculated to bring the cow into a state in which she would stand serenely to be milked. I began to wonder how many farmers were cruel and stupid.

At the end of the year of Agriculture I had, at least, experienced enough botany and zoology to know that I wanted to become a biologist.

Victory against Japan had come during that year, and the university now had many students who were ex-service personnel. Some were resuming studies they had begun before the war, but many were taking advantage of their veterans' benefits to gain an education they could never have afforded before. The gap between those of us coming straight from school and these older students seemed great in years and maturity and were, of course, unbridgeable in terms of experience of life. Yet, a year or two into our course work, many of us had formed permanent friendships that ignored differences in age and wartime experience.

By the time I began my second year in biology about fifty students, the majority of them girls, were heading to graduation in biology, and nearly all of us were taking zoology, physiology and genetics.

In Sydney University at the time, those studying physiology took a course that was two-thirds physiology and one-third biochemistry, both taught by the medical faculty and with much the same material as given to medical students with whom we shared lectures and many lab exercises. Most of the female biology students would become schoolteachers, dieticians or hospital technicians. The men were nearly all hoping to become professional biologists. This split had nothing to do with relative academic ability. Women did as well as men in their studies, often better. Yet there was an inextinguishably optimistic feeling among the men that they could somehow make careers in biology.

In those early post-war times, Australian society was buoyant with expectations of forging a future in a country that, to most people, had begun to seem of infinite promise. Indeed, as students of the sciences graduated, the expanding economy continued for years to create a nation-wide job market that somehow accommodated nearly everyone who wanted to work in science. Why most who wanted this were men rather than women can only be explained as a phenomenon of the times. Speaking for myself and many other males who were graduating in science, they were lucky times for people with such meagre experience and training as was ours.

At the same time, Sydney University was extremely poor financially, overcrowded, with a creaking infrastructure, a small and ageing teaching staff, ill-equipped, ancient laboratories and obsolescent libraries. Little research was going on and there were very few graduate students. The place was in the Australian post-WWII doldrums along with most of the country.

Nevertheless, many of us found the lectures in zoology, human physiology and genetics absorbing. This could not be said of many of the laboratory exercises in zoology that consisted of laborious dissections of formalin-preserved stingrays, rabbits, crayfish and, for some highly obscure reason, a detailed study and dissection of the anatomy of the sheep's brain.

To balance the examination of this and other dead material in hot and stuffy classroom labs, we enjoyed a total of two days in the open. On one of these field days we viewed the plant succession in the massive sand hills behind one of Sydney's largest beaches. On the other field day we examined the plants and animals in rock pools below a sandstone headland. The latter was for me the transformation of a childhood wonder into the beginning of science. These were brief but exciting and informative occasions during which we drank in the sights and habitats of scores of wild and living species.

Looking back, it seems incredible that no formal courses in ecology were offered to biology students—these day-excursions were the nearest approach we experienced. How bizarre it would seem a few years later. We could have learned so much, so rapidly, in a few days in the open in visits to shores, lakes and forested mountain areas, all within easy reach of Sydney. What could trying to match our dissections to the idealized drawings in books and manuals tell us that would be of the slightest later use to us? It was as if we were training to be doctors or butchers. Yet somehow, despite the absurdities and waste of time, the university seemed to promise an abundance of intellectual riches to come ... somewhere down the line.

My first real hint of how and why sustained research of any kind could be exciting came not from zoology but from human physiology. The head of the department of physiology was Professor Frank Cotton. Of the thousand or more students in medicine, pharmacology, veterinary science, dentistry and general science who studied in the courses of his department, the numbers were divided into two groups of similar size. One was taught by a physician-physiologist, the other by Frank Cotton himself. Their styles, aims and method of presenting the material differed in nearly every way.

I believe that most of those taught by the physician thought themselves fortunate. His lectures were compact, well constructed, easy to relate to the standard textbook and to make notes from. As for Frank Cotton, he simply addressed himself in each lecture to some matter that had arisen in physiological research, or he told us how some great researcher had made progress, or he concerned himself with the "first principles" that lay at the heart of all science and of our ability to understand the world in scientific terms. He also tried to make clear to us how vital it was to study magnitudes and measurements and numbers if we were to make sense of scientific data we had collected. He infuriated many people, because of their desperate entreaties to tell them what was "important" for their success in examinations, Frank, with maddening geniality, would say, "It's all there in the textbook, all you need to know. It's a very fine textbook; read it. We will expect nothing from you in the strictly factual sense that is not more than adequately covered in the book. But I am not here merely to read the book to you. I have to try to get you to think for yourselves." Yes, many hated him.

To me, he was, at the time, the epitome of what a scientist must be. I learned more from him in the way it affected all my views of science than from all other teachers over all the years. He was not a prolific producer of research publications in the modern mould. But he was looking for major achievement that would change the face of the area of science in which he was active. His field was the rational application of the knowledge of exercise physiology towards maximization of performance in sports. He was the first Australian researcher to introduce the principles and practice of "interval training" to athletic endeavours. Frank Cotton was able to exercise great influence on actual progress in these sports, enhancing and optimizing the performance of both individuals and those in teams. By the adroit use of specialized testing machines that he designed with great facility and insight he was able to measure the effects of training regimes with a hitherto unavailable accuracy, advise athletes how to bring their performance potential to a maximum and then retain their competitive edge without injury. He also showed them precisely how to apply their abilities in the most effective way on the day of competition. I will recount a concrete example.

Edwin Carr was a medical student at Sydney University. He was also an Australian champion runner at 400 m, and later a Commonwealth Games champion. He was a great athlete, though he did not remain active in the sport long enough to tell whether he could become a truly international star. Frank Cotton advised him in the use of interval training and he was one of the first great Australian athletes to use this method. It had been standard practice for middle-distance runners in Australia to run several miles at very slow speeds with occasional outbursts of sprinting for 50 or 100 m or so. Yet, before WWII, interval training had been used by some athletes in other countries with great success.

Woldem Gerschker who was one of the founders of this training method had advised Rudolf Harbig in the method. Harbig set a great world record for the 800 m of 1 min 46.6 sec and also a record for the 400 m. He was killed in WWII.

Frank Cotton knew of all this and strongly recommended the use of interval training to Australian runners and swimmers. I remember Edwin Carr as he reeled off successive laps with one lap at 80 per cent of his racing speed to be followed by a lap at a slow canter, sometimes even a walk. He might cover twenty of these pairs of laps in a series. He would add a few dashes of about 50 to 100 m in sprints. At this time many Australian runners and their coaches believed this regimen was absurd, bound to exhaust and injure the athlete. Cotton insisted it would build endurance and speed. And he was right.

In 1949 the reigning 400 m world record holder, Herb McKenly of Jamaica, visited Australia and the local media were agog with his prowess and how far behind our local stars were. Frank Cotton saw an unparalleled opportunity to kindle public interest in athletics and in the merits of science-based training methods.

A race track was prepared for a contest of McKenly against Sydney runners on a 400m course at Sydney Cricket Ground. The city had no quality cinder tracks and the course was grass and circular. Cotton knew that McKenly was not used to running on grass or on a circular track that lacked a finishing straight. He also noted that McKenly, running at his fastest, would normally be expected to head a runner of Carr's quality by up to 10 to 20 m at the halfway point. McKenly usually ran the first 200 m at the incredible time of less than 21 seconds and then hung on to win, even though the rest of the runners might be closing on him at the end.

"All right," said Cotton to Carr. "Let him go. But at the halfway point start to close on him and run the last 100 m as if your life depended on it. He will already have a huge oxygen debt and actually be struggling—which is the pattern he has adopted. He has enormous ability, but from the physiological standpoint he does not run a rational race. You will not reach your maximum oxygen debt till very near the finish. You may get very close to him at the end."

And that was how things went. A large crowd attended to see the great McKenly show the local lads how it was done. Carr closed the gap between them at the end and ran out winner by a metre or two. The time was "slow" on this grass track without straights, and Carr collapsed for a minute while the Jamaican just panted a bit. But it was a win of incredible portent for Australian athletics.

And I admired Frank Cotton enormously. He had logically analyzed the effects on McKenly of the way in which he ran, pointed out his tactical weakness and addressed how, on cold physiological grounds, a less experienced and less gifted runner might defeat him, particularly on a surface he was not used to. It was a masterly lesson in the application of common sense and scientific principles in the solution of a challenging problem. It happened to be in sport, but it taught me all kinds of lessons I hoped I would never forget if I became a scientist.

Cotton's eventual influence on training athletes was huge, and many Australian athletes of the late 1940s and the 1950s and far later owed much of their great performances in international competitions to his teachings and advice. It all resulted in helping the country attain the athletic fame it acquired from 1945 and on into the future.

Frank Cotton showed that a combination of sharp thinking based on established scientific knowledge, coupled with cleverly conducted, simple experiments, using test apparatus he designed, could yield measurable, replicable and highly significant results. By these means he was able to change much of the understanding of a field of applied science. I suppose it was strange that I, a future zoologist, would take fire mostly from a human physiologist, but the lessons from Cotton's ways of viewing and doing science were dramatic and full of suggestive ideas. Nor was he the man to disregard a significant detail because as he said, "Straws show which way the current flows."

In my final undergraduate year in zoology a young scientist, who had already earned some reputation as an insect population analyst, joined the zoology faculty. This man, Charles Birch from South Australia, had studied the population dynamics of thrips and weevils under experimental conditions. He had also visited the great ecological strongholds of the English-speaking world—the university research communities of Charles Elton at Oxford and the animal ecology group at Chicago. These people had been among the most prominent leaders of ecological work and thinking for a generation and Birch had benefited from his experience with them. But he did not come to us as an ecologist. What we learned from him was comparative animal physiology and something of the thinking of those philosophers who were asking how living organisms had originated, how human consciousness had arisen, and how, and whether, animals regulated their own population numbers. Birch was a meticulous and ingenious experimenter and interpreter of data. I found him fascinating but rather stern in manner and have always regretted that I did not get to know him better—perhaps in graduate studies. What I did learn from Birch certainly played an important part in how I came to view the animal world.

In the following chapters it will, I hope, become clear how the ecological work I did subsequently extended my education and made me gradually aware of the nature of ecosystems and of the tasks and challenges animals face and encounter in their struggles to survive as they complete their life cycles. Biologists, like other scientists, never "complete" their education. Nor do they usually become encyclopaedias of scientific information. They become instead people who learn how to investigate and better understand those aspects of Nature that interest them.

CHAPTER 2

MOUND ANTS

After graduating I managed to obtain a research fellowship in physiology. I remained in this appointment for two years, long enough to help Frank Cotton in some brief experiments in training and testing athletes. It gave me further glimpses of what his works and insights in exercise physiology were leading to in the world of athletics. It also afforded me the chance to meet and observe some great athletes, but it did not lead to a possible career. For that I would have needed a medical degree and my resources ruled that out. Instead, I turned back to my fascination with ants and worked on them whenever possible.

I was living in Thornleigh, a northern Sydney suburb on the edge of a large bush reserve, and began to study there the behaviour of the mound (or meat) ant, *Iridomirmex detectus*, the very same dark red-brown species I had watched and played around with as a child.

This ant species was very common, often raiding garbage cans and invading houses to carry off fragments of spilled of food. Mound ants could make their nests wherever there was enough soil, so they were able to inhabit city gardens and make small nests in the soil below cracks in paving or crevices between rocks. In places where space and soil were plentiful and their movements undisturbed, mound ants could construct and occupy nest mounds that ranged in size from a few centimetres to several metres across.

Ants, bees and wasps are all members of the scientific Order Hymenoptera, and those species of this order that make their homes in dense colonies are termed the Social Hymenoptera. They are remarkable animals. For various species of bees it has been shown that they are divided into castes, and among the tasks they perform are locating, and signalling to other bees, the location of food, as is

scientifically and now popularly well known. The Austrian biologist
Karl Von Frisch was awarded a Nobel Prize for his studies of the
behavioural ecology of honeybees. Many ant species are just as
interesting in their general biology as bees, if not more so.

Many species of fish, birds and mammals are also highly social
animals, but they are vertebrates, all with individually different and
complex nervous systems, and their individuality can usually be
recognized. One rabbit is not exactly like all other rabbits in its
behaviour, breeding biology or even its personal physiology. And any
particular rabbit's method of struggling to stay alive will differ
however slightly from those of other rabbits. But if one considers a
single worker ant, no matter how interesting its behaviour or
physiology, or even its anatomical features, may be, they really only
have meaning in relation to the ant society of which it is a
functioning unit. For instance, the death of an individual ant in
defence of the nest has no more significance than that of any other
worker. The death of a single worker will not be the death of an
actual or potential parent of the ant society. Only in connection with
the survival of the few specialized reproductive forms in a nest or
colony does the survival of workers take on much significance for the
ant society of which they are members. Likewise, most workers do
not feed themselves as an end in itself, whether or not they are living
within a social unit, but instead carry food to their co-workers and
those ants of other castes, or immature stages, that are in the colony.
The differences from the vertebrate social animals are not trivial.
Bees and ants really can only express themselves as units of a society
that is intrinsically far more complex in its emergent properties and
functions than are those of the creatures that compose it. Indeed,
many biologists have found it useful to view colonies of social insects
as "superorganisms," higher units in their functioning and biological
significance than the simpler organisms that comprise them.

Mound ants attracted me as objects of investigation because of
their abundance and because these fairly small insects were
collectively able to manipulate large volumes of soil, food and other
materials for their use even though their activities as individuals
appeared so simple. I decided to study them from an autecological
standpoint. This meant that I would concentrate on this single
species as its members acted with respect to its environment, which
would include determining how it responded to various conditions

and other creatures it encountered. Thus I would be trying to arrive at an ecological characterization of the species.

I began by deciding to observe the more conspicuous activities of the ants on several large nests for some days to obtain a very general impression of their behaviour and to give me a starting point. I hoped such a starting point would enable me to begin to make sense of their otherwise almost impenetrable activities. In considering the ecology of all animals one must begin by identifying some aspects of their biology that seem major and prominent in their behaviour and ecology. Your decision may prove less significant than you originally think it is, but your choice will remove the confusion of not knowing where to start, and will anyway open the way to other aspects of lives that may eventually prove significant. Choices of how to begin must be made in beginning all ecological studies, otherwise confusion will rule in contemplating the apparently depthless complexity of the life of organisms.

It seemed, almost immediately, that studying the ant traffic leaving and entering the nest mounds at different points and at different times of the day would be an interesting way to begin. I mean that it was obvious that these ants were leaving the nest to search for sources of food or materials for mound building or to remove dead ants or other wastes, and that many could be seen returning later, some bearing food or other materials such as pebbles or tiny twigs. Many ants left and entered the nests along trails, though not all did so. But it was most convenient to count ants moving in the trails near the nest where numbers were at their greatest. Numbers leaving fell off with distance along the trails as ants dispersed in their searches, so that at last it was no longer possible to identify trails as numbers using them approached zero.

Ants were active in leaving the nests along trails from first daylight hours, and the counts of ants crossing lines at right angles to trails close to nests were made for six-minute periods at various times during the daylight; ants did not leave or enter nests by night, though they could be active on nest surfaces. Most of the diurnal activity curves made from the data showed a tendency for activity to peak in early to mid-morning and again from mid- to later afternoon. Apart from the zero activity levels in trails at the beginnings and ends of daylight hours, the lowest activity roughly coincided with the days' highest temperatures and lowest relative humidities. Although the main trails emanating from particular nests remained constant in position, they were not all used every day. Despite some correlation between temperature, humidity and activity in the trails, it appeared likely that the basic tendency towards bimodal activity peaks was endogenous. Activity got going quickly early in the day as most of the searching ants left the nest, then dropped away, and finally picked up again as the bulk of the searchers came rushing back to the nest towards evening, bringing with them twigs, pieces of food or perhaps nothing visible.

The orientation of mound ants as they moved either away from or towards the nest was difficult to reverse. Stiff notepaper placed in front of a moving ant would cause it to reverse its movement for a moment; then it would turn back again. With repeats of movement blockage, the ants could eventually be turned. Ants returning to the nest were harder to turn than those moving away from it.

Scraping away strips of soil, even when up to a metre wide and 2 cm deep, failed to prevent the ants from following a trail. Nor was their trail-following affected after heavy rain. Some ants certainly follow odour trails but it seemed unlikely that this ability, if present, was much employed by mound ants.

Mounds were generally capped with fine gravel, though cemented soil crumbs or small twigs, even charcoal fragments from past fires, could also be used. The body of the mound was formed of excavated soil particles brought to the surface by the ants. In the placement of these particles the ants would tend to move rather randomly, the dome shape of the mound apparently being mostly the result of rain and gravity. Sectioned mounds revealed a series of galleries of irregular shape and disposition both in the mound and below ground level. Some galleries opened directly on the mound surfaces.

Whereas many single or isolated social animals (particularly mammals) can act with a high degree of coordinated and cooperative, yet flexible, behaviour in performing various tasks (lions and wolves are good examples), the activities of social hymenoptera should be interpreted with caution. Though many ecologists and ethologists have remarked on the extraordinary integration among them, in mound ants, moving materials to, from or across the surface of nests, or in the trails, was sometimes the result of the actions of single ants, at other times of several. The direction in which small groups of foraging ants moved objects depended on the forces resultant to the different strengths and directions of the pulling efforts of different ants in the group. In most cases there appeared to be little coordinated activity, except in the sense that several ants pulled at once. This kind of activity may be common among ants; some investigations have shown that when the respective forces by each of a pair of ants pulling on one object were measured and the resultant force was small it was possible that the individual efforts of the two ants were so applied that only a small proportion of their efforts was contributing to the resultant force and direction of movement of the object. The remainder of the forces were wasted in the opposition between the two. Larger resultant forces, with more effective and rapid movement of objects, were more likely to occur only after ants had been pulling for eight to ten minutes. These observations do not paint mound ants as very intelligent or spectacularly efficient in their cooperative efforts! Any sense of "purposefulness" in their joint activity should be viewed with caution.

Defence of the nest adds further to this picture. When I placed another invertebrate such as a small millipede among ants active on a mound surface, an elevation in ant "excitement" was certainly observable in the millipede's immediate location, and this excitement both intensified and spread as I repeatedly turned the millipede back from the edge of the nest. But it never made it impossible for the millipede to escape, even when hundreds of excited ants stood in its way.

The excitement elicited on stamping on a mound, or thrusting a stick into the surface, was a similar type of thing: a wave of activity that would be propagated across the entire surface of the nest but which was most intense at the site of damage. With the bigger nests, the waves would be very weak, or negligible, on the parts of the mound farthest from the stimulus.

When the mound surface was not stimulated there was still almost continuous ant activity on it during the daylight hours. Some ants would emerge briefly then re-enter the nest; others wandered apparently at random. These random movements seemed inexplicable. Of course, there would always be some entering a trail to join its traffic leaving the nest.

Ants returning to the nest might bring specific small pieces of food in their jaws but it appeared that most of the food carried would be sap exudate from trees for regurgitation feeding of ants in the nest. Objects such as twigs were carried often quickly but awkwardly by single ants for deposition on the mound surface. On the whole, single ants appeared more efficient and "purposeful" in carrying objects than did two or more ants!

The fact that several nests existed near each other raised the question of how independent from each other their inhabitants were. Ants were removed from each of four nests and then replaced on their nest of origin or on another nest. The ants from "other" nests, though not attacked, were examined carefully by the ants of the receiving nest. One trail that seemed common to two nests had many examples of ants in which aggregations up to about twenty strong occurred along the course of a common trail. These gatherings were generally circular in shape, with heads more often towards the centre, but the forms were somewhat plastic. The bodies of the ants were raised on extended legs, their abdomens elevated, and movements were jerky and sudden; quivering was almost constant. Pairs would touch antennae, then break apart, perhaps to repeat the performance with others. The duration of this behaviour was variable. Again, some rejection occurred of "transferred" ants.

Let me note at this point that conservationists frequently use this type of autecological approach in studying a rare or endangered species, but if they are investigating an animal species it is most likely to be a vertebrate: a bird, mammal or fish. I had been doing it on a very abundant and ubiquitous, small, invertebrate species. However, even if mound ants were not, and never would become, rare animals of great public interest, the approach I employed was the same as it would have been if they were a more "interesting" species.

If what has been recounted had been all that could be said about the mound ant we might summarize as follows. The mound ant is an intensely social species whose principal daily activities are performed entirely by unspecialized workers (male or female alate, that is sexually active individuals, were not seen in the period of this study). Workers construct dome-shaped nests of a wide range of sizes, the surfaces of which were perforated by simple round openings connecting with the galleries in the mound and below ground level. Workers left the nest in their daily search for food and building materials in maximum numbers in the morning and returned in maximum numbers towards the end of daylight. They might follow permanent or semi-permanent trails whether leaving or returning, but could travel unaccompanied. The trail-following was not easily disrupted even when major damage was inflicted on the trail. The ants were active either in the field or on the nest surface during the daytime, but only on the surface at night. The ants moved objects either alone or in groups of up to about a dozen, but there was little sign of highly coordinated or efficient activity; many objects got moved towards or away from the nest, or on its surface, because of a resultant of forces rather than a cooperative effort. Defence of the nest took the form of excitement at the site of injury or damage and spread as a wave across the nest. The excited ants would swarm and bite, but at least to this human the bites were innocuous. They were incapable of immobilizing fast-moving prey, even when the prey was small. Ants from different nests would stand together on trails in some kind of "confrontational" small groups. From all appearances the mound ant was a very abundant species, of markedly social but unspecialized behaviour, very unlikely to be easily threatened by other species or by climatic changes.

"Well and good" one might have said: a series of simple facts gleaned in a short-term investigation of this species. A longer study would have yielded far more detailed and more quantified information about the species, such as the age of nests in relation to their size, number of ants present in a nest, the establishment of new nests following nuptial flights, etc.

The work was published in 1953. Twenty years later, R. D. Hughes published a summary of a very long-term, but formerly unpublished, study of the species that had been performed by Tom Greaves in his book, *Living Insects*.

Greaves' work had been done around Canberra, the Australian Federal Capital. In the early days of Canberra the landscape was an area of sheep pastures that lacked mound ants. They became established following much tree planting during Canberra's development. Greaves found that among the larger nests there were many instances when small satellite nests were established so that a connected network of nests of many ages and sizes (up to 10 m across) eventually covered large areas.

The behaviour of mound ants in these areas resembled what I had found in Thornleigh, with a few differences. Thus the nest traffic in Canberra was almost entirely confined to trails, whereas I saw a good deal of movement to and from the nests by individual ants as well. Behaviour on the nest surface was much the same, as was the continual coating of the surface with protective materials, and in Canberra, the ants kept the surfaces clear of plants, which was not needed at Thornleigh where the conditions were very arid with meagre groundcover.

Nest traffic in Canberra was similar to that in Thornleigh, except that trails in Canberra were much more visible because of having been worn through pasture grass. It is also claimed in Greaves' work that the ants lay down a trail-following pheromone and that a pheromone also accounts for the waves of activity that accompany attacks to the nest surface. Actually, an alarm substance released by mound ants was identified in the 1960s as methyl heptanone by a scientist in the University of New South Walesa scientist in the University of New South Wales identified an alarm substance released by mound ants in the 1960s as methyl heptanone. The Canberra ants seem to have been more aggressive on the trails and on the nest mounds, but differences of this order may have been results of many local environmental influences. As with the Thornleigh ants the Canberra ones were opportunistic foragers and also sought the sweet exudates of trees. The confrontational postures of ants in shared trails seen in Thornleigh had also been found in Canberra by Greaves and, with the perspective of his long-term studies of mound ants, he was able to assign this behaviour to situations where separation of nests formerly connected was about to occur.

Up to this point, then, the elements of behaviour and ecology found in the brief study at Thornleigh were quite similar to those found in Greaves' work. But in his very long study he was able to unravel the major features of the entire life cycle of the species.

Greaves was able to excavate and examine nests and identify males and females that grew to larger sizes than ordinary workers and had the potential to become winged and sexually active. They undergo their final moults in late summer and the virgin females leave the nest and subsequently the males also emerge from the nest. The males fly high and later the females fly off as conditions warm. This they continue to do while warmth continues, though they will retreat into the nest if the temperature cools. The major part of Greaves' long study describes the subsequent death of drones and reproductive life of females, the positioning of the larval ants in the mounds and the size, age and numbers of ants per nest in relation to nest size. These are fascinating facts but go far beyond the scope of the present chapter.

It was important to me that though neither Greaves nor I could build on the studies of others, our findings (setting aside the differences that might be attributed to their physical environments) were in good agreement. Greaves worked on mound ants for 20 years; my work was only for a few months. My studies were, of course, slight compared to his. But I felt satisfied I had at least asked many of the questions Greaves did, and found similar answers. This was reassuring, because sound science depends on repeatability of results. In our case, neither of us knew anything of the work of the other.

CHAPTER 3

A WORLD OF ECOSYSTEMS: LEARNING THROUGH MISTAKES

An ecosystem can be defined as a system formed by the interaction of a community of organisms with their environment. It follows that the environment cannot be taken merely as the organisms' containing physical shell of climate, soil, water and so on, and indeed for organisms in a particular locality their environment will also include other organisms. A community is defined as an assemblage of plants and animals in a given space. Obviously these two terms—ecosystem and community—are closely related and are often used almost interchangeably.

For all scientists there are particular ways in which they view the world. The world view of most ecologists is dominated by their tendency to think of Earth as a planet of ecosystems. Because all ecosystems have their roots in the past it may be necessary, in considering any present ecosystem, to try to understand what preceded it and predict, or attempt to infer, what it may someday become.

Effective ecologists and conservationists should always have in mind that ecosystems are populated by living organisms, always be able to imagine or know them in a pictorial or schematic sense. Ecologists must also continually employ numbers, quantities and symbols in their ecosystem problem solving. They must not forget they are examining the life cycles and conditions of existence of real microorganisms, real plants and real animals.

I don't recall any lecturer mentioning the word "ecosystem" before 1948 when I was a university undergraduate in Australia. True, we went on expeditions to seashores and coastal sand hills, but I don't remember use of the term ecosystem with reference to those settings. Again, in 1950, when I was searching for employment in either wildlife or fisheries research, there was still precious little, if any, talk about ecosystems, and really nothing about conservation. But how could biologists throughout Australia be working on problems that were clearly ecological, with obvious conservation implications, avoid using these terms? In other countries several well-regarded scientific journals were almost wholly dedicated to these topics, as were many textbooks. Why was this not so in Australia? Probably because most ecologists were employed to work on how to eradicate or drastically control rabbits and so-called pest insects en masse and on how to maximize the yields of fisheries, forests and grain harvests. Conservation was barely mentioned. With the vastness of Australia compared to its very low human population of no more than eight million, governments of the day were assuring that conservation, except for water for crops and drinking, could hardly be a serious problem.

Well, though I thought a bit about ecosystems and conservation as I began to read widely in the literature of ecology, it was not until I began to work in fishery research as a member of The Commonwealth Scientific and Industrial Research Organization (CSIRO) in the State of Tasmania that both these concepts began to hit home.

My first task in Tasmania, beginning in 1951, was to take over and evaluate the results of a field experiment that had been started by others in 1948, and had been largely abandoned because of insufficient personnel to complete it. It had been begun in the following way: A senior aquatic biologist in the Division had been greatly interested in experiments carried out on sea-lochs in Scotland during WWII. It had been at a time when normal North Sea marine fisheries, a very important part of the country's protein food supply for centuries, had been suspended. A group of marine biologists and hydrographers, hoping to make a contribution to wartime food, had hit on the idea of boosting fish production in sea-lochs (elongated stretches of seawater open to the sea only by narrow channels). There were plenty of accessible sea-lochs that contained wild fish populations in some British coastal localities.

Thinking in ecosystem terms, the biologists argued that the environmental conditions in sea-lochs were already known to some extent, together with their fish species and the various organisms the fish ate, so that it might therefore be feasible to introduce chemical fertilizers into lochs to stimulate their productivity. By this means they might be able to augment the food chains to edible fish and significantly increase their growth rate and production. They chose to work on two Scottish sea-lochs: Loch Craiglin, only 18 acres in area, and Kyle Scotnish, a mile long, both of them connected to the sea by a much larger area of seawater, Sailean More.

In 1942, work began on Loch Craiglin, to the water of which were added totals of 319 pounds of agricultural superphosphate and 515 pounds of sodium nitrate in a series of constant fractions of these amounts. Similar amounts were added in 1943, and a smaller amount in 1944. The loch was fertilized a total of eleven times in 20 months.

The microflagellates of the phytoplankton were the first organisms to respond, showing increases two to three days after each fertilizer addition at all depths. But if utilization was rapid, the increases were of brief duration. The increases among the larger phytoplankton organisms were delayed until the autumn when the densities of the diatoms and dinoflagellates exceeded those in the outside Sailean More. Zooplankton organisms, such as rotifers and larval molluscs, also increased over their numbers in Sailean More. Among the bottom-living (benthic) invertebrates, increases were delayed for a year, but then there were massive increases.

The loch had previously lacked flounders but in 1942, 2,600 small flounders were released and their subsequent growth exceeded that of flounders of similar age in Loch Killisport, from which they had come, by four times in length and sixteen times in weight from July 1942 to April 1943. Soon after, 1,100 more flounders were released in Loch Craiglin and in the summer more than 21,000 young-of-the-year flounders were released. All of these three stocks of flounders grew very much faster than in Loch Killisport. On the face of it, this experiment in an attempt to increase fish growth looked highly successful but, for several reasons, too detailed to be gone into here, the extremely rapid growth in Loch Craiglin may not have been quite as directly related to chemical enrichment as at first seemed certain. The second experiment, in the greatly larger Kyle Scotnish, played out in a fairly similar way, though this loch was under direct marine tidal influence.

Both these Scottish sea-loch experiments drew much attention, despite some uncertainty in the results. Some of the design faults were attributable to wartime exigencies.

My work on the effects of chemical enrichment of a Tasmanian lake began early in 1951 on Lake Dobson, a small (7 ha) oligotrophic lake of glacial origin at 1,000 m altitude in a dolerite mountain region in Mount Field National Park. It was about 50 km from Hobart. My responsibility was to work as a research scientist on "investigations of freshwater fisheries potential." My knowledge of freshwater biology was practically nil. Nobody would be directing or supervising my work and I would have to train myself to do the required job. With no more than a B.Sc. to my name, the proposition would seem ridiculous today, but it was the common experience of a host of young Australian scientists of the time. You sank or swam through your own efforts, or by how luck favoured you.

A couple of days after arrival in Hobart I was driven to Lake Dobson, our field vehicle climbing on a gravel road for the last few kilometres, past spectacular waterfalls and huge manferns and traversing a dark, cool-temperate rain forest of sombre trees that stood like sentinels. We arrived at a high, narrow, glacial valley, a wild, stark landscape of rocks overlain by a skeletal soil that managed to nourish dark, heathy shrubs. And everywhere there were snowgums with their smooth, pale, twisted limbs. This was a primal Nature, tough, grand, looking as it would have before humans came. It was a zone of winter influence, of endurance. The air was pure and clear, stingingly cold though it was only autumn. It told me even before I began to examine Lake Dobson that my prescribed job was only going to be one aspect of the experiences that were coming. As I stood, shivering but elated, I felt that, if ever, it was now, here, that I was going to become an ecologist.

Lake Dobson itself was a rocky cupful of cold water, with a very faint amber colouration from the peaty qualities of its drainage basin. When I dipped up a glass jar of its water I saw just how utterly clear it was, with a natural crystalline sparkle I could scarcely have imagined. The brownish-yellow I thought I had been seeing in it as I gazed into the lake's depths had been that of the colour of the mud deposited in its basin, and which was clearly visible through seven metres of transparent water.

This day was overcast, but on another day, clear, fine and cloudless, the sky hard blue, like faultless but unreflecting glass, I climbed the steep walking track that traversed the adjacent mountainside and looked down at Lake Dobson, three hundred metres below. The water sparkled up at me, a great eye, blue as lapis lazuli in the sunlight.

It was in this Mount Field National Park, that contained Lake Dobson and other lakes and tarns (including the larger Lake Fenton, which was the main Hobart water supply), that I first came to appreciate words like "wilderness" and "conservation," first felt the timeless sanctity of Nature and the interrelatedness of living things.

When I first saw Lake Dobson, and for at least one or two subsequent years, I was still in that state of essentially unconfounded loyalty to science that is more common to its younger practitioners. That is, I would applaud most of the things of which science appeared capable, except in the making of weapons of violence and war. I mean that I found I could marvel at what had been done to Lake Dobson through manipulating a natural lake for a "practical" purpose, even though it was obvious that this had profoundly violated the lake's pristine integrity. I know that as we crossed the lake in a dinghy I was amazed to see great growths of an aquatic plant, the water milfoil with the scientific name *Myriophyllum elatinoides*. The plants were a rich deep green, with whorls of tiny leaflets. They grew, evenly spaced in four lines that crossed the lake, two lines following the lake's long axis, two at right angles to these. Each line ran from shore to shore. Most of the plants reached the water surface, which means that the largest were not only one and a half metres wide, but also were seven metres tall in the lake's deepest part. They were like richly-leafed young trees growing clear to the surface. This was when my views of science first became conflicted. For what I was trying to absorb seemed shocking to me more than it was impressive. I mean, inexperienced in viewing Nature as I then was, it was quite clear that, in the midst of an unsullied, purer environment, it was as though I had come upon a vast, artificially created garden. Even to my inexperienced eyes this was simply "not on."

Back in Hobart it was explained to me that when observations first began on Lake Dobson in 1948, three years before, it had very little of either phytoplankton or zooplankton and the growths of milfoil were small and scattered. This was the result of its extremely low mineral content. I learned that the chemical enrichment intended to increase production in the lake began a year and a half later in August 1949 when paper bags filled with 4.6 kg of a mixture of agricultural fertilizers (20 parts ammonium sulphate, 60 parts superphosphate, 5 parts potassium chloride, 15 parts ground limestone, a mixture that yielded an abundance of nitrogen, phosphorus and potassium) had been dropped in lines from a rowing boat two years earlier with the hope that they would penetrate the lake's soft bottom mud. The reasoning was that the paper bags would sit in place and release the fertilizer slowly and sparingly instead of it going into solution rather rapidly and being mostly lost to the lake via its outgoing stream. It was thought that this could provide a long-lasting nutrient supply for the growth of plankton, and dependent aquatic microfauna that would in turn mean an enhanced food chain and food supply for fish. But the bags had sunk more deeply into the mud than expected and much of their contents appeared to have been trapped below the surface. Enough was released to promote a phytoplankton bloom, mainly of diatoms during the next few months, but the main result was vast nutrient enrichment of the mud in the vicinity of the bags that caused the giant resultant growths of milfoil. No one had anticipated so rapid and prodigious a growth of these plants, small specimens of which had already been in the lake as part of its natural bottom flora. It was like a Jack-and-beanstalk story! Zooplankton catches remained very low, however.

After another year and a half, starting in October 1950, the same total amount of fertilizer that had been in the bags was broadcast over the lake from a rowing boat in a series of seven applications. During these applications zooplankton began to increase and over 1951 its abundance reached a level at least 2,000 times greater than its former value.

Let me be clear: the experiment had been botched in many ways, badly planned and poorly carried out. I was thankful none of this was my fault. However, what I now had to do was to try to interpret the results as well as possible. And in doing this I had to teach myself a lot of biology, and I learned first-hand just what the term "ecosystem" really meant.

I began to examine the earlier data files on Lake Dobson. Water analyses had been suspended and needed to be restarted immediately, as did plankton catches and estimates of the abundance of organisms on the milfoil (which had never been done before). Later, the effects of the fertilizer additions, if any, on the growth of trout that made their home in the lake and spawned in the rocky outlet stream would have to be assessed. I was face to face with a research problem involving a whole lake ecosystem.

I soon found out that some of the people from my CSIRO divisional headquarters at Cronulla in Sydney had built a sampling device for collecting mud from around and among the root systems of the large milfoil growths. They had established that more than a year after the paper bags penetrated the mud most of the fertilizer chemicals (phosphorus, nitrogen and potassium) remained in the roots about a foot below the mud surface. To see if the huge growths of milfoil could really be directly associated with increased quantities of minerals in their tissues I collected sample plant material from both the giant growths and from other milfoil plants that were little affected by the paper bag experiment. All the nutrient elements were at higher concentrations in the tissues of the giant plants, especially phosphorus, suggesting that, even if the nutrients at their roots were gradually being diminished, the plants themselves might continue to grow for a very long time.

The natural condition of Lake Dobson had been that of a very oligotrophic lake, that is, very unproductive. This low productivity was determined by the surrounding landscape of dolerite, a very hard, insoluble rock. Early analysis of the water of the lake had shown it to be very low in dissolved chemicals, including nutrients such as phosphorus. In fact there was so little total mineral content in the water that it could probably have been used in a car battery.

The records showed that the earlier additions of fertilizer had produced phytoplankton blooms, but these blooms had not been properly recorded or measured. The blooms of phytoplankton abundance had been, in particular, short-lived (as in the Loch Craiglin experiment). Given the lake's natural biology it is likely such blooms had not occurred in it at any time since the last glaciation, ten thousand years earlier. Understanding this simple likelihood added a scope and depth to my thinking about what had so rashly been done to Lake Dobson's long-time environmental integrity.

Zooplankton—microscopic crustaceans (copepods)—had also undergone a population explosion. The latest phytoplankton bloom was now over but zooplankton remained abundant compared to its amount before enrichment. What was the source of its food that had not been available before? I was not able to prove it, but concluded that the ongoing dying and decay of the now-abundant milfoil was providing plentiful particulate food for the zooplankton.

The milfoil was also the source of other food and shelter for freshwater animals. I had a Hobart metalworker build me a stainless steel grab for seizing intact samples of milfoil together with the organisms it harboured. A plentiful fauna, dominated by stonefly nymphs and aquatic gastropods (small snail-like molluscs) was living on the huge plants. These would be prime targets as food for the trout in Lake Dobson. And now there was even more milfoil than before, for additional rich growths had grown diffusely around the giant growths since the second, broadcast applications of fertilizer.

It had not originally been known for certain that there were any trout in Lake Dobson. In 1952, 467 tagged yearling brown trout were released in the lake after chemical enrichment had produced its main effects. Thirty were later recaptured and my senior colleague A.G. Nicholls assessed by direct measurement and by back-calculation from growth rings on their scales their growth while at liberty in the lake. But it was his back-calculation of the growth of unmarked fish that had already been in the lake before the enrichment began that proved more interesting. Scales of sixteen such fish showed clear increase in their growth that coincided with the period of chemical enrichment. This effect was more notable because recaptures of the marked fish in 1954 showed that these had come to comprise about two-thirds of the total trout population in the lake. This meant that increase in growth rate of resident trout occurred despite a large increase in size of the original population. So the experiment did answer its major point: fish did apparently avail themselves of the great increase in food supply that occurred as a result of the milfoil growth.

And at the end I really started to understand what an ecosystem was and how it worked. The plants (in the form of phytoplankton and clumps of higher vegetation) lay at the base of food production in the lake, and invertebrate animals in the form of zooplankton or insect epiphytes lived on them. Fish lay at the next (third) level in the "trophic" system. The particular ecosystem that was Lake Dobson had basically low productivity (or growth)—

whether of plankton, plants or fish—that had been elevated to a much higher trophic level by the nutrients in the form of chemical fertilizers which had been put in the lake. I am here, of course, simplifying the nature of the lake ecosystem, omitting, for one thing, any references to microbe-driven cycles of life and breakdown functions in water and mud. On the other hand, the fertilizers did demonstrate that enrichment of the lake water by appropriate chemicals did produce large plankton "blooms" and masses of higher plants with their associated insect and mollusc faunas, and that these events coincided with increased growth of fish. So a real feeling of the dynamic production processes in the lake emerged, and I could truly see the lake as an ecosystem that responded powerfully to a nutrient level increase without having its original main organisms destroyed or replaced. It changed its productivity but not, in this case, its character. I should also note that from its level of about 80 per cent of oxygen saturation in 1950, a rise in oxygen occurred in the lake's water to 100 per cent or more, during the period of chemical enrichment and great growth of milfoil. This was the result of increased photosynthesis by the now much greater amount of plant material in the lake.

Things can work out strangely. I had found something in the confusion and mess of the Lake Dobson experimental setting, which should only have originated after a full preparatory study of the lake ecosystem before any attempts had been made at moving an oligotrophic highland lake in the direction of eutrophy to make it more suitable for rapid fish growth. But, anyway, what was the real point of the work? Its basis was the interesting but flawed experiments on Scottish sea-lochs. But that work was at least the result of wartime emergencies. Lake Dobson was in post-war Australia, a highland freshwater lake that could be frozen for months at a time, which was unrecognized as a resort for fishing and whose primary "function" was as a base for the people who liked to walk on rough trails through Tasmanian highlands.

Nevertheless, serendipity (always to be watched for in science) had been at work. I had been forced to study Lake Dobson at a limnological ecosystem level which had resulted in my coming to understand this in a functional way that opened my eyes to all ecosystems. The lessons had been swift, and I had learned them alone while on the job. But they were not the less for this.

Besides, another serendipitous circumstance lay in just coming to Tasmania. Tasmania was a place where landscapes were overwhelming, impenetrable, lost to the world. Even 50 years ago, so many places in this planet were already being revamped, moulded according to human will. Much less so in Tasmania. Its natural landscapes, whether cool temperate rain forests, high altitude heaths and lake lands, or clean, empty beaches and rocky coasts, were uniquely manifest. It ought to have been beautiful, but I do not think beautiful is the right word. In Tasmania the solitudes and silences inspired a sense of awe and grandeur. You could stand alone, waiting, in a place that was very old, unknowable, different from other places. But for me there was often a feeling of presences watching from the deep scrub; presences that were about to emerge quietly but who thought better of it whenever you turned and might have greeted them. It was not a sense of something threatening, rather of creatures that had been hurt too much and now were very cautious. I was continually reminded that genocide had been practised against the native Tasmanians. The last of the full-bloods, the few survivors of earlier days when they had been hunted almost to extinction, died just before 1900.

In many ways, I loved Tasmania. But I always asked myself, for all the wisdom it imparted in the way that Nature lived and worked, if there was a curse on the place.

CHAPTER 4

INTRODUCED FISH: RECREATIONAL INTENT; SCIENTIFIC SPINOFF

A major challenge for ecologists and conservationists is to determine the principal factors that influence or govern the distribution and abundance of plants and animals. This may be a far from easy problem even when a species under investigation is native to its environment. But when one is considering an exotic (i.e. introduced) species, the problem may include extra degrees of difficulty. After its introduction considerable time may need to pass before a species expands into what will be its eventual living area. The rate and extent of spread may depend on the number released and their ability to locate a suitable reproductive habitat. The first releases may be eaten by native species or may contract a pathogen from them. Climatic conditions may happen to be exceptionally unsuitable in the years of release and cause major mortalities in environments that would otherwise be inhabitable.

On the other hand, the performance of exotics in their new environment can be of great interest in providing further critical tests of the validity of opinions about the factors that are supposed to limit them in their home environments. Thus, Australia has been a valuable proving ground for these aspects of field ecology.

During the 19th century, anglers who had learned their skills in Britain seemed unable to resist bringing to Australia species of fish with which they were already familiar in their homeland. This was not unique to Australia, of course. How else did trout end up as established populations in the highlands of Malaysia and Mount Kenya, equatorial locations in which trout survived in waters that were high enough to remain cool? Australia has not only introduced fish but also rabbits, pigs, cattle, horses, sheep, goats and bird species (including poultry); early explorers brought in camels as pack animals. All are in wild populations. There are hosts of plant species.

My friend, the late John Lake, and I set out in the 1960s to consider the distribution and abundance of introduced freshwater fish in Australian inland waters, specifically the Murray-Darling River system. Our working position was to try to determine whether the conditions of their Australian occurrences resembled those where they were native. The main problem we set ourselves was to plot the distributions and ranges of four introduced species (brown trout, rainbow trout, European perch and tench) in relation to general habitat and climatic conditions. All these species now have their major occurrences in self-reproducing populations in the waters of south eastern Australia, specifically in the Murray-Darling River system which flows west and south from the Eastern Highlands and, in the case of trout, also in streams that flow from the Highlands east towards the Pacific.

Trout

Brown trout (*Salmo trutta*) were brought to Australia in 1864 and, via Victoria, reached their main distribution and range in New South Wales in 1888. In 1894 rainbow trout (*S. gairdneri*) were brought to New South Wales from New Zealand and thence to Victoria and Tasmania. Lake and I were fully aware that biologists living in the northern hemisphere would automatically accept that trout occurrences there were associated with (controlled by) water temperatures and other stream habitat conditions necessary for spawning. We also knew that trout populations had been successfully established in places such as the tropical highlands of Malaysia and Africa, where water temperature conditions are cool enough and essentially constant throughout the entire year. Our question was, then, whether the surviving trout populations in the Australian mainland would show similar locational limitations, that is, be confined to the highlands but otherwise able to flourish. Our conclusion was that in New South Wales and Victoria brown and rainbow trout do in fact survive as reproducing populations in the tableland streams of the Eastern Highlands, but mostly above 600 m altitude.

However, based on the direct experience of John Lake more than 40 years ago, or information conveyed to him at that time, we reported that trout could also be found in streams down to 300 m altitude when the gradient from higher altitudes was steep enough to deliver cool water rapidly to lower altitudes. It also appeared that, in some instances, water at lower altitudes was kept cooler than it would otherwise have been by discharges from large storage dams. This let trout survive at unusually low altitudes. However, conditions in most of the trout waters below 1,200 m in the central and northern highlands of New South Wales were too hot for trout, or at best marginal, in some summers. The only apparent exceptions to the absence of trout at low altitudes at that time of the year appeared to be in streams flowing east to the sea, in which their presence was probably the result of small numbers being washed downstream, even into estuaries, as results of severe flooding in higher streams.

Trout also occurred in the lower reaches of the great river systems that flow west from the Eastern Highlands across New South Wales. For instance, both species were sometimes caught along the course of the Murrumbidgee River and also in the Murray River in places beyond the South Australian border, and even into the Darling River. To reach such locations trout would have to have descended from higher altitudes, for temperatures that can reach or exceed 30°C in summer, would otherwise prevent their presence. Even though these trout were able to survive for some time in river environments well downstream from their main occurrences, the absence of the clean gravel beds required for their spawning redds would have made reproduction very unlikely.

On the whole, then, we found that trout were limited to the same kinds of environments in the mainland of Australia as they are to those in the northern hemisphere. There were no great surprises; their distribution could have been as predicted in advance of their introduction. The "secret" of their success in south eastern Australian rivers, which can be summed up in a few words, is that they are largely in areas at relatively high altitudes where water is not too warm for them and there are plenty of stream substrate conditions suitable for spawning.

I must add that we reported that also in Tasmania the two trout species tended to occupy habitats of types similar to those in the northern hemisphere, but their main distribution patterns were influenced by the efforts of humans. Many Tasmanian rivers contained trout along the whole course from small elevated headstreams to estuaries (i.e. sea level). If some seemingly suitable systems lacked trout we believed it was probably because they had not been stocked. Trout existed as self-maintaining populations in both small and large Tasmanian lakes, given that they had conditions for spawning. High summer temperatures appeared too low to set limits to trout ranges in Tasmania, nor were there many rivers that lacked enough gravel to prevent trout from finding spawning substrates. Trout were also able to survive in small water bodies such as farm ponds, but again, could not persist there unless they had access to spawning areas.

In Europe, trout tend to be thought of as inhabiting the higher reaches of rivers that appear too slow downstream where the rivers are mainly given over to other fish. But this picture did not apply well in Australian streams. This was probably because the idea of dividing streams into zones characterized by their fish species occurrences at various levels was based on the big European rivers with relatively even and gradual progression from highland to estuarine conditions. It has actually been pointed out that the European concept of fish distribution is not even directly applicable to British rivers which tend to be comparatively short and irregular in their courses.

It also seems possible that, in the diversified fish faunas of the large European rivers, fish encountered some degree of interspecific interaction that helps to shape the structure of the communities. By contrast, in Tasmanian waters, only three native fish species achieve significant size: the eel (*Anguilla australis*), the blackfish (*Gadopsis marmoratus*) and the sandfish (*Pseudaphrirtis urvillii*). These species are carnivorous and may compete with trout for food, but apart from eels, their numbers and biomasses tend to be small. Members of the family Galaxiidae, indigenous to Tasmania and with many species, can inhabit streams with trout, but are generally too small and few to offer trout important competition for food or space. Galaxiids, however, certainly serve as food for trout, so their presence may help trout to attain rapid growth and large size.

Studies on trout in New South Wales by John Lake found trout growth rates that were decidedly, in some cases spectacularly, greater than for trout in Tasmanian, New Zealand and British waters. There may be various ways of attempting to account for this high growth rate, e.g., the trout originally released may have been from very fast-growing parent stocks; the trout may have undergone some kind of genetic selection for fast growth in the Australian environment; there could be special qualities in the Australian aquatic environment including rich food supplies. However, we found that the length of the annual period of temperatures most favourable for rapid growth was clearly greater in New South Wales than in the Tasmanian, New Zealand or British environments. On the basis of the available evidence it seemed scientifically soundest to adopt this, the simplest, interpretation.

Tench and perch

Tench (*Tinca tinca*) were brought to Tasmania in the 19th century; tench populations were firmly established in certain rivers by 1882. They bred successfully, but were restricted to the sorts of slow, weedy rivers that offer the same kinds of shelter and spawning areas which they favour in Europe.

European perch (*Perca fluviatilis*) were introduced into Tasmania in 1862. They reached Victoria as an original release of seven in 1868 and were introduced into New South Wales in 1888. Although tench and perch are not at all closely related (tench are members of the Family Cyprinidae, perch are of the Serranidae), the essential patterns of their life cycles resemble each other's.

Both species favour slow-moving and weedy streams and, compared to trout, were slow in establishing themselves over large areas, since they are not free-swimming and in the case of the tench are very cryptic, not easily seen unless actively hunted or trapped. Neither species has a spawning run, and both deposit their eggs among weeds or on other substrates that offer support and shelter. Their distributions in Australia reflect their European habitats and distribution.

In Tasmania, which is the smallest Australian State, the spread of tench has placed them in approximately the same sorts of river and lake environments as on the mainland. In some of their river habitats that are close to the sea, their salinity tolerance was shown by some of my studies not to be great enough to allow them to move along coasts to colonize many additional streams that could otherwise prove suitable. Neither have tench penetrated highland regions, except in a few places to which they have been deliberately introduced. This failure is presumably because they are not "strong" swimmers, and avoid the faster-flowing streams they would have to traverse to reach the highlands unassisted. In this, their distribution parallels that in the northern hemisphere.

The picture for both species is fairly similar in the Murray-Darling waters of the Australian mainland. To recapitulate, these species are unable to penetrate the highlands, except when deliberately released there, and again their occurrence is in weedy and slow-moving waterways. Here, the tench were found inside the range of perch; they had not spread out as far from their original site of release in approximately the same time period, presumably because they were even more slow-moving and cryptic in their behaviour. Neither species had crossed the highlands to enter rivers running east to the sea.

There is one more interesting point about perch range. Experimental and zoogeographic studies that I carried out showed that perch cannot survive in temperatures above 30 C. In the Darling River, which extends deeply northwards into Queensland, we found that perch could not extend beyond a point at which summer temperatures exceed 30 C. This thermal barrier to their range was as real for this species in both Europe and Australia as are fast-flowing waters, or some other impassable physical barrier.

Notes on reproduction

In many ways the most interesting, yet little remarked aspect of the lives of freshwater fish species in Australia—and an aspect that occurs all over the world where species have survived being moved from the northern hemisphere to the southern—is that the transferred species are able to adjust the timing of their reproduction to match the calendar reversal of the seasons. Thus, of freshwater fish species that evolved in the northern hemisphere and were successfully introduced into Australia (including four species not specifically considered here), six originally reproduced in the northern hemisphere in spring to later summer. In Australia they do the same, which means they are, in calendar terms, approximately half a year out of phase. It certainly seems remarkable that it is not more frequently noted. After all, it demonstrates, though it is well known that the successful timing of the breeding part of the life cycles of these fish species are tied importantly to certain features of the yearly environmental cycle, that these processes are not very hard-wired by evolution. That is, the timing of these important phases, as they exist in the northern hemisphere, and where their breeding is considered as closely limited by seasonal light and temperature, have the ability to become readily and rapidly adjusted to the new calendar timing of the environmental conditions of the southern hemisphere. In other words, the timing of the reproductive life of these fishes is quite facultative.

Thus these results in the huge field settings of an alien environment demonstrate important ecological aspects of the lives of fish species, with a certainty that experiments in the limited environmental settings of laboratories or hatcheries would find it is almost impossible to match.

Updating the picture

Recently, Dr. Mark Lintermans of the University of Canberra has presented work on the range and distribution of the four fish species in the Murray-Darling system that Lake and I worked on some forty years ago. Scientists are always interested in seeing how well their studies have stood the test of time and this can be very important for ecologists since changes in climate, inter-specific interactions, changes in food availability and the like can bring about significant modifications. Besides, our working hypothesis had been that the final distribution of these fish in Australia should have been essentially predictable from what was known of them in the northern hemisphere. Forty years ago we could say that as far as we could see, the essential picture of the ranges of the brown and rainbow trout and the perch had probably reached their limits in the Murray-Darling system and that this would also apply to the perch. We did state, however, that the range of tench, which was then within that of perch, would be likely to expand over time and come eventually to occupy much the same range. Dr. Linterman's maps show that this is essentially what has been happening. Perch are still apparently about where they were, with tench still within the total range of perch, but now only just.

Thus, even though there are many more recorded distribution sightings now than 40 years ago, trout and perch have not changed much in their distribution and range, except that tench have increased their range considerably but that it still lies within that of perch.

The moral of this tale is that ecologists, and especially those who are hoping that their assessments of environmental limitations (here the introduction of alien species into an environment which had contained only indigenous species) are correct and will stand the test of time, are naturally relieved when this works out.

I do not want to ignore the fact that where fish species have been introduced into Australian freshwater environments they may frequently have done serious, perhaps irreparable, harm to native species. This has come in many ways, in the case of trout from competition for space and breeding territory and for food, to serious impacts of trout as predators on smaller and vulnerable native species. In fact, the real point of the results is that they represent a failure to apply the knowledge that in many cases already existed about prospective introductions and that would have led to fairly accurate predictions of unfavourable consequences. In this account I am trying to point out that what happened is a very good example of where factual knowledge did exist, but was not applied. In a sense one can attribute this failure to the fact that ecology was not yet recognized as a science in the 19th century. Today the very term "ecology" implies the recognition and naming of collections of facts as a subject is often to lend them a power they had not formerly possessed.

These examples should serve to emphasize that when people consider altering environments either by changing physical conditions, or introducing new plants and animals into localities where they did not formerly exist, it is worth taking the time and trouble to attempt to predict what the short-term and long-term results may be. Such an approach should neither be ignored nor underestimated. For instance, what we have really learned is that as a result of our combined knowledge from the distributions and ranges and physiological tolerances of some species, we can pretty well define where they can live in various parts of the world. We can be sure that as the globe warms the ranges of many fish species will be inexorably redefined. Fish already sensitive to high temperatures will be driven to higher altitudes and latitudes. Trout in Australia will eventually come to inhabit fewer and narrower areas. In the long run they may find the waters in which they have done well are no longer tolerable to them.

Let me add as a tailpiece a note not on fish but on perhaps the greatest piece of biological blundering to which Australia was ever subjected. It is known widely and famously that Australia has for a century been overrun by introduced European rabbits. The rabbits, in their untold millions, have been a blight on the pastoral industry, and have contributed to the degradation of the soils and landscape. They have been hunted, shot, trapped, poisoned and infected with the myxoma virus—all costly of money, time and effort. Yet if ecological thought had been applied to their introduction, all this need never have happened, because they would not have been allowed in.

Senseless to mention this? Perhaps. But consider this. Well back into the early 19th century a host of people—farmers, hunters, gamekeepers, amateur naturalists—already knew these things about rabbits. They were widely distributed over Britain and France but their numbers in wild populations were kept in check by predation of foxes, hawks and eagles, by hunting, poisoning, use of ferrets and dogs, etc. They could live in climatic conditions ranging from warm temperate to cool, but not thrive in extreme cold or heat. Rabbits released in small numbers by mariners on oceanic islands that lacked predators had soon overrun these islands. After all, rabbits breed like … rabbits!

Australia had enormous space, few predators, very little intensive farming and a southern half that contained huge areas where the climatic conditions should favour rabbits. Given all that was already known about rabbits why was there not enormous opposition to their introduction? I suggest it was because the word for, and concept of, "ecology" did not yet exist. A name makes a difference, because it signifies there is a recognizable body of knowledge that can be discussed, debated, referred to and used to make hypotheses and projects, ideas and principles.

Today, ecologists can often advise people on how to avoid their most egregious and disastrous errors about Nature and its creatures.

CHAPTER 5

THE MOLONGLO RIVER AND LAKE BURLEY GRIFFIN

The problem and its setting

This is about an ecological study of the Molonglo River and its implication for Lake Burley Griffin. The river was polluted by zinc, a particularly important matter as it would be supplying water for an artificial lake that would lie at the heart of Canberra, the Australian Federal Capital. So, in 1960, it was hoped to obtain the ecological understanding that would lead to suggestions for the management and conservation of the lake. The investigation turned out to be a rather extensive example of the use of the toolkit of ecology in studying a conservation problem. However, when it came time to hope for, even expect, some applications of the findings that could lead to significant conservation measures for the river-lake system, there were some problems. Nevertheless, the work provides an example of the necessary connection between ecology and conservation, an example of how ecologists initiate, develop and perform their analytical views of the world that can lead to rational conservation practices.

In 1908, it was decided that Canberra, the future federal capital of Australia, would be deliberately located in an area in the pastoral highlands of the State of New South Wales, designated as Australian Capital Territory. The choice of this site was a compromise, since the two largest cities in Australia, Sydney and Melbourne, were vying to be chosen as the Federal capital. With a flair, unexpected for the young and thinly populated nation of Australia, the initiators decided to hold an international competition for the design of what would be the city of Canberra (whose name may derive from a native term for "meeting place."). The 137 design plans submitted displayed much ingenuity and imagination and showed 3D concepts for the buildings and landscape of the city that ranged widely from neo-classical to modern. The winning plan, in 1912, by a noted American architect, Walter Burley Griffin, was based on a bold scheme. The Molonglo River flowed through what would become the centre of Canberra and it sometimes rose alarmingly following heavy rains. As finally conceived Canberra would be a "garden city" with an artificial lake consisting of three connected basins which would be filled by the Molonglo River. The lake's shape would largely result from the topography of the site. The lake would serve as a flood control reservoir and also add to the city's recreational and aesthetic aspects. Under most conditions the lake's depth would be controlled by an adjustable outflow as part of the lake's dam that could be opened as required.

When I joined the Zoology Department faculty at the Australian National University in 1960 the university had already been requested to determine whether the amount of zinc in the Molonglo (zinc is toxic to fish) would be too high for the survival of rainbow and brown trout in Lake Burley Griffin (named for its architect). Before 1938 the Molonglo River had been a good quality trout angling stream and many anticipated that if the water in the lake proved tolerable some angling would be possible.

Looked at critically, however, it seems remarkable that a lake which would become a major landscape feature of Australia's capital could be permitted to be filled with metal-polluted water. However, the Australian Capital Territory exists geographically within the State of New South Wales, and the Lake George Mine (the source of the zinc) was in New South Wales. The mine had been in operation since 1882 and did not fully close until 1962 when its operation became uneconomic. It had provided miners' jobs for many of the inhabitants of the town of Captains Flat. But on the mine's closure, no environmental clean-up was demanded of the Lake George Mine owners.

The task accepted by the university, and for which I found myself responsible during the years 1961-64, centered around an evaluation of the chemical and biological conditions in the Molonglo downstream of the mine, and was to determine the nature and severity of the pollution.

In beginning this work I invited Dr. J. R. Beevers, a biogeochemist of the Commonwealth Bureau of Mineral Resources, to join me, and we were later joined by (now Professor) P. S. Lake who was at the time an Honours student at the university. Our two main questions were whether fish, and the invertebrate animals on which they feed, were able to survive in the polluted Molonglo.

At the beginning we had to concentrate on the physical entity and characteristics of the Molonglo River itself, which rises in the Eastern Highlands of New South Wales, about 50 km SSE of the city of Canberra at about 900 m altitude. The river passes, with small tributaries, through country that is composed of areas of granite, greywacke shale, limestone-bearing shale, acid volcanics, gravel and silt. It was clear we needed to review the past history of the effects of the mine and also carefully examine the catchment of the Molonglo watershed in the vicinity of Captains Flat, where zinc, in the water from the abandoned mine, and from its wastes, entered the river.

The Lake George Mine had been closing down its operations only a couple of years before our work began. At first sight of its straight main street, battered buildings, dust and abandonment, I thought immediately of a film set for a western movie. The effect was heightened by the abandoned mine equipment, rusting rail lines and the massive piles of mine waste (slimes dumps). The old-fashioned, still functional town pub lent an additional air of seeming authenticity to the environment and this was even more powerfully reinforced by the discovery that the pub's bar was one of the longest in the country! There had once been a lot of thirsty miners there.

And there was a further air of quaintness and eccentricity. I asked the origin of the name "Captain's Flat," half anticipating to hear a story of some retired military or naval aristocrat long settled in the countryside. "Oh, Captain?" someone said. "That was the name of an old horse that was kept in a field in the town. Everyone knew it." Field ecologists can have many interesting experiences in addition to those of a strictly scientific kind!

It might be assumed that since a multitude of experimental data already existed on the effects of zinc and other toxic metals on fish, measurements of the amounts of zinc in the river before the completion of the dam that would allow the lake to form would be routine, and would readily lead to indications of the possibilities of fish surviving in the polluted lake. But nearly all the evidence concerning zinc toxicity was for fish and other aquatic organisms came from laboratory studies in which zinc had been set at particular levels and temperatures. The data on fish survivorship, though voluminous, were thus essentially useless to us, since we would be dealing with wide, order-of-magnitude differences of zinc in the Molonglo as its concentration responded to changes in rainfall and stream-flow conditions. So we largely abandoned trying to characterize the course of the changing severity of pollution in the river in chemical terms and resorted to alternative approaches, of which the main ones were to concentrate on studying the invertebrate animals (termed the macrobenthic community) that populated the river bottom, and to test the survival of fish directly in the polluted Molonglo mainstream. Our reasoning was that the river benthic fauna was constantly exposed to zinc, so that their variety and numbers would reflect the gradual changes in the impact of zinc, the average concentration of which was assumed to decrease with distance downstream from the pollution source. This profile would show us where, if at all, conditions reverted to normal. The approach itself was based on the pioneering book by H. B. Hynes, *The Biology of Polluted Waters*. And what we were saying to ourselves and others was that the presence of zinc in the river in measurable amounts was our essential, established starting point. After that it was best to see directly what happened to organisms exposed to it at increasing distances downstream.

We also performed chemical studies in attempting to understand the geochemical mechanisms that caused zinc pollution.

The necessary background

About 11 km northwest of its source, the Molonglo River passes through the village of Captains Flat. Virtually all the pollution in the river was primarily or secondarily traceable to this site, and because the nature of this pollution in its geochemical and biological manifestations was the subject of our study, I give here a statement of its character.

Mineralization was discovered at Captains Flat in 1874 and mining began in 1882. Discontinuous mining followed and ceased about 1900, and the mine was not re-opened until 1924 when the original deposit was reappraised. The mine was in full, continuous operation from 1938 until closure in 1962 and it was in this period that the major pollution of the Molonglo occurred. The earlier activities were evidently minor in impact on the Molonglo which was, as stated earlier, a good trout habitat before 1938. But in 1939 a slimes dump, resulting from the deposition of mine waste, collapsed and acid water with high concentrations of copper, lead, iron and zinc entered the Molonglo River. This mishap, however, was evidently not a major one. In 1943, there was another collapse as a result of which an estimated 30,000 cubic metres of fine tailings consisting of pyrite (FeS), and siliceous material containing galena (PbS), chalcopyrite (CuFeS) and sphalerite (ZnS) slipped into the reservoir holding the domestic water supply for Captains Flat. Water equal in volume to the collapsed dump entered the Molonglo and finely divided zinc tailings plus, presumably, some copper and lead were conveyed many miles downstream by the flood. At Captains Flat the river was blue and further downstream a light bright green. The methods of chemical analysis then available precluded a more detailed knowledge of the events, but the gross observations imply that the river carried a large base-metal load and was probably extremely acidic. However, the pH of the river was reportedly approaching neutrality only ten days after massive pollution.

The sudden rush of water following the slimes dump collapse caused extensive flooding, starting about 15 km downstream from Captains Flat in the region of Foxlow Plain. There was widespread destruction of vegetation, and this devastated area remained, the effects still chemically detectable in the early 1960s in river bank soil about 50 km downstream from the collapse.

After 1943, acid mine water was pumped into the Molonglo and drainage water from the massive waste dumps also entered the river, both conveyed via the small Copper Creek.

Altogether, the pollution that reached the Molonglo River was acidic, from base metals in solution, and from very finely divided tailings material scattered along the course of the river and in suspension in the water. Thiocyanates, sulphides, cresols, phenols, etc., which were used in the ore treatment processes, were never a serious threat far downstream from the mine. They usually met acid mine water before reaching the Molonglo, and sulphides and cyanides were removed as their volatile acids. Phenols and cresols were removed from the water by oxidation. At the time of our investigation we concluded that ongoing pollution of the Molonglo was coming from the already deposited tailings material.

Before the mine closure, dissolved zinc showed the expected tendency for high values immediately below the pollution source with progressive general reduction in values with distance downstream.

This, then, was the basic setting of the knowledge that had accumulated at the beginning of our work. After the mine closure our chemical studies indicated that zinc concentrations in the river were derived from the disposal stacks of mine waste and mine water at Captains Flat, supplemented by oxidation of finely divided sphalerite material from the earlier spills being deposited down the course of the river and its shores. More on the details of these processes is given later in this chapter.

Though the mine operation was for copper and lead in addition to zinc, concentrations of the former two elements appeared very slight in the course of the river.

We added zinc sulphate to each of eight 10-litre samples of Molonglo River water taken from just above the junction with the unpolluted Queanbeyan River to make concentrations ranging from 1.5 to 51.5 ppm, and allowed them to stand in glass jars uncovered in the laboratory for seven weeks. Zinc concentrations determined for each sample at intervals during the course of the seven weeks showed decreases of zinc to about one-fifth of the starting concentrations. We did this experiment because we already had suggestive evidence from field observations that concentrations of zinc were being lowered along the course of the river by hydrolysis in addition to the effects of dilution from inputs by unpolluted tributary streams. By comparison, when zinc sulphate was added to Canberra tap water, no zinc was lost from experimental samples over similar time. In other words, waters of differing chemical contents may behave with marked differences according to their zinc loads.

As the above experiment was proceeding to add to our understanding of the environmental setting for the organisms that survived in this chronically zinc-polluted system, we established a series of stations in the Molonglo River at which we could examine the benthic animals along the course of the river.

The ecological entry point

As I mentioned before, finding ways into an ecological problem is always the essential requirement. For us, this now meant characterization of the invertebrates (the macrobenthic animals) that were surviving in the river. However, we realized that the presence or absence of some groups of these animals could be influenced by environmental conditions other than zinc levels at each station. The station sampling went like this: the three unpolluted "control" stations were in the Molonglo itself just above the pollution from the mine, in the unpolluted Whiskers Creek, and in the Queanbeyan River (Stations 1, 8 and 12), which formed the stations at which we expected to find a normal invertebrate benthic fauna for this river system. They would provide a baseline picture against which we could view whatever damage the presence of zinc in the river was doing to organisms in the Molonglo. The polluted series downstream from Captains Flat were Stations 2, 3, 4, 5, 6, 7, 9, 10, 11, and 13.

Special mention should be made of certain stations, starting with the unpolluted ones. As for Station 1, it was immediately below Captains Flat dam that held back the town's water supply. Here the river was swift, turbulent, shallow and apparently unpolluted. It ran over an uneven but stable rocky bed. The banks were lined with tea-tree (*Leptospermum* spp), and sedge (*Carex appresea*), and on the rocks aquatic macrophytes and algae were abundant in the water.

Station 8 was in Whiskers Creek, a small, unpolluted tributary of the Molonglo, which entered it near Station 7. It had slow flow and few riffles, a bottom of small pebbles and clay, and abundant algae and macrophytes.

Station 12 was in the unpolluted Queanbeyan River about 1 km above the junction with the Molonglo and below a small weir at a caravan park. Below the weir was a large turbulent pool followed by a wide shallow riffle. A stable bed was of medium-sized stones and gravel with abundant algae. Slight intermittent organic pollution actually did reach it from the town of Queanbeyan.

As for the polluted stations, Station 2 was directly below Captains Flat, where the stream was swift, polluted and the banks largely devoid of vegetation except for a few scattered patches of reeds (*Phragmites communis*).

Algae were few and the stream bed was of medium-sized to large stones and unstable stretches of gravel.

Station 3 was in Copper Creek, the main source of pollution from the mine wastes. The creek was slow, the bottom of large stones between unstable banks of loose gravel and soft clay. The bottom was covered with an opaque milky-brown deposit, apparently of ferric hydroxide. Algae were absent.

At Station 4 in the Molonglo, just below where Copper Creek flowed into it, algae were almost absent, the banks were devoid of plants and the stream bed was largely unstable gravel.

What should be noted of these three stations, which were the most heavily and directly polluted, is that they were also characterized by lack of vegetation and very unstable bottom conditions. Farther downstream, at Station 5, where the Molonglo ran across Foxlow Plain, the river was broad and shallow and both algae and macrophytes were still scanty except in quieter and deeper stretches. But though banks were somewhat unstable and eroded, as was the bed of sand and gravel, conditions were less unfavourable than in the preceding four stations. As for the remaining stations, it was not until Station 11 was reached that stable conditions in the riffles were found.

The polluted stations displayed considerable variation in conditions so far as stream bed and abundance or scarcity of algae or macrophytes were concerned. Station 13, considerably the farthest downstream from Captains Flat, was closest to the eastern end of Lake Burley Griffin, with whatever remained as slight pollution from the Molonglo at that point. The river was slow and deep with only a few short riffles. Turbidity was always high because of gravel washing operations a few hundred metres upstream.

All stations featured some riffle areas, vital if satisfactory sampling comparisons were to be made.

The river invertebrates

We were now in a position to examine the macrobenthic faunal communities at the stations in the Molonglo and in the three control situations: one in the upper Molonglo and those at Whiskers Creek and the lower Queanbeyan River. The simple collecting techniques we employed are still in use in contemporary freshwater ecology.

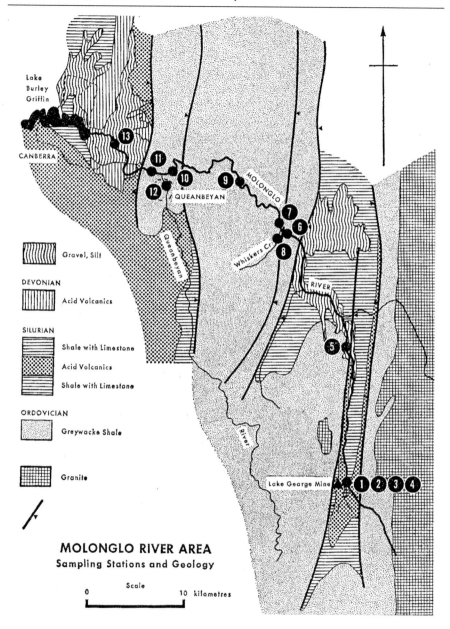

Figure 1 Molonglo River showing sampling stations and geology. (From A. H. Weatherley, J. R Beevers and P. S. Lake (1967), *The Ecology of a Zinc-polluted River*)

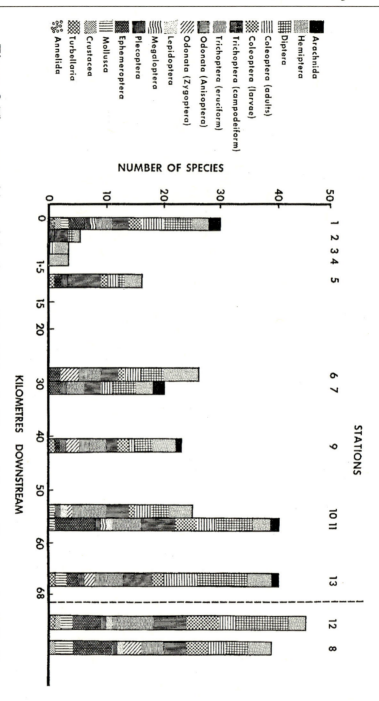

Figure 2 The composition of the fauna of the Molonglo River and "control" streams in 1963. (From A. H. Weatherley, J. R Beevers and P. S. Lake (1967) *The Ecology of a Zinc-polluted River*)

All stations were sampled by collecting as many animals as possible from a 0.1 sq m quadrat sampler with an attached, small-mesh net. This unit allowed the collector to "wash" all stones, gravel, sand and mud that lay within the sampler's square frame, and the flowing water swept animals thus dislodged into the attached collecting net that hung downstream. This method is quantitative as long as the current is sufficient to ensure that all animals dislodged are carried into the net opening and not carried over or around it. Long-handled sampling nets of nylon mesh on a rectangular steel frame were also used. By collecting for a standard time and always with an identical net and in a similar manner, much additional information was obtained that made it easier to compare the fauna at the various stations and to establish the ranges and types of species present. During each five minutes of collection time about a hundred metres of stream length were sampled, mainly along edges and shallows.

Before considering the results of the collections of invertebrates we noted some physical and chemical conditions in the Molonglo that might have influenced the diversity of the species found.

The main length of the river was slightly alkaline, approximately pH 7.4. Of course, under certain conditions pH values below 5 or even pH 4 in Copper Creek were recorded. Oxygen levels were always adequate or high, and free carbon dioxide tended to increase somewhat with distance downstream from Station 1 and the same trend attended calcium, magnesium and conductivity values. The Queanbeyan River was lower in hardness and conductivity than the Molonglo. However, it was only in the case of Copper Creek that very low pH and high conductivity, in themselves, could have been expected to have significant noxious effects on the macrobenthos.

In a river such as the Molonglo, though the entire system below Captains Flat was polluted by zinc, there were also great increases of short duration in zinc values due to rainfall. At some points in the river, zinc was mainly in the form of zinc sulphate. At other points it was in the form of finely divided tailings from the flood plain deposits at Foxlow that were mainly laid down in the 1943 flood. To study the complexities of this situation in more detail would have required the collection of vastly greater data sets than we had the resources to obtain. Almost certainly, the most toxic effects of zinc on any fish in the river would likely have been short-lived, and far above the highest concentrations of zinc that would have been enough to eventually kill fish over extended periods. This is why, once having established the environmental presence of zinc, we deemed it simplest and best to employ the resulting patterns of diversity and abundance of animals in estimating the overall "severity" of its toxic effects. Similar reasoning was applied to experiments with fish in live-boxes. Our idea was that if fish were exposed to the river conditions at selected locations for an appropriate time their survivorship would reveal just how unfavourable conditions of pollution were at different parts of the system.

Figure 2 shows the total number of invertebrate animal species of different groups caught by hand net sampling along the course of the Molonglo and in the unpolluted river above Captains Flat and in Whiskers Creek and the Queanbeyan River. Note the extremely small number of species in Stations 2, 3 and 4, immediately downstream from the sites of maximum pollution compared to Station 1 (immediately above the pollution zone). Further downstream, however, at Stations 6 and 7, the number and range of species were again becoming similar to those at Station 1, and of course in the results obtained from Whiskers Creek and the Queanbeyan. Though not shown here, the abundance of animals collected in the 0.1 sq m quadrat sampler paralleled the number of species. The numbers were lowest just below the pollution source, then abundance rose to maxima at Stations 10, 11 and 13. The majority of the few species found in the highly polluted Stations 2, 3 and 4 were essentially restricted to the adult stages of species of aquatic Hemiptera and Coleoptera (bugs and beetles), which are both air breathers with chitinous body coverings that are water-impenetrable.

If one wants to go further in understanding the effects of heavy metals, especially zinc, in influencing a river fauna several factors can be noted. Zinc toxicity is inversely related to water hardness. This fact may help explain why the species diversity at Station 11 was markedly greater than at the very nearby (but upstream) Station 10, because between them the river passes over a limestone band. Others have found that what might be considered a mildly zinc-polluted stream would likely lack crustaceans, molluscs and oligochaetes (worms). This is the picture of the Molonglo upstream of Station 10 to Station 2.

There is also the principal indirect effect of zinc that can be expected to have broad ecosystem implications that will reduce the quality and stability of the aquatic environment as a whole. Particularly in riffles, where the scouring effects of running water are highest, algae and macrophytes tend to hold the stream bed stable. If polluting zinc kills them off, the gravel and coarse sand of the stream bed become comparatively loose. This may result in damage or mortality to the stream fauna. When we look back on the conditions at the various stations it was clear the stream beds at Stations 2 to 6 were unstable, and indeed that it was not until Station 11 was reached that stable bottom conditions in the riffles were found.

Fish

Now came field experiments with the survivorship of live fish in the Molonglo. For many years the Molonglo between Captains Flat and the junction with the Queanbeyan had been spoken of as "fish-free." That may never have been completely true. Certainly there is a vast amount of evidence that even low concentrations of zinc are toxic to fish. But both the Queanbeyan River and the zinc-free tributary of Whiskers Creek contained small numbers of trout in self-maintaining populations. Some fish from Whiskers Creek may occasionally have moved into the Molonglo. If so, they could not have survived there for long. A dam on the Queanbeyan River probably prevented its fish from penetrating the Molonglo pollution region.

We tested fish in the Molonglo by holding them in wood-framed boxes covered by nylon mesh secured in the mainstream at four stations: two of them just below the zone of heavy zinc-pollution, a third, about 30 km downstream (where the macrobenthic animal communities were about halfway towards normalization), and the fourth at a locality where the macorbenthics were largely at normal values.

In writing these notes I can recall—still with some amusement—how my friend and field assistant at that time (now a recently retired professor at Cornell University) and I were concerned that the live boxes, once installed in the river, might be discovered and interfered with. My friend was both young and occasionally impetuous, and took the initiative of attaching notices to the boxes to the effect that their contents were radioactive and potentially hazardous. I needed to point to him that the notices, if seen, might well bring down on our heads attention vastly more damaging to us and our cause than mere field damage to our experiment! He accepted my judgement, though at first not without some argument. In any event, we took great care to conceal the presence of the boxes as much as possible.

Over a sixteen-day period of exposure to the field conditions, the brown trout demonstrated survivorship that was superior at all stations to that by the rainbows. In fact, at the station farthest downstream from Captains Flat all the brown trout survived compared to 60 per cent mortality of the rainbows. Two additional experiments employed wild carp and catfish, not because either species would be of interest to anglers, but merely to add the variety of two more species of different biological characteristics and sensitivities to zinc to the mix. The survival pattern of the carp roughly resembled that of the rainbow trout, but all the catfish succumbed within a week.

Broadly speaking, the results of the fish mortalities corresponded with those for the benthic animals. Thus, of the brown trout, all died in less than 16 days in the two most polluted sites (Stations 4 and 5), while 60 per cent had died by day 16 at Station 6, and at Station 10, the next downstream and least polluted, all had survived. The survival pattern was similar for the rainbow trout exposed at the same time and stations, except that fish died off more rapidly, and even at Station 10, 80 per cent were dead by day 16. The second experiment gave somewhat similar results; brown trout again outperformed rainbow trout, though this time 40 per cent had died at Station 10 by day 16 and no rainbows survived beyond 12 days at any station. So brown trout were more resistant than rainbows to the levels of pollution in the Molonglo and, even though toxicity did decrease with distance downstream, it appeared unlikely that fish could survive for long anywhere along the river's course. To judge from our experiments, hydrolysis would have been one of the means by which zinc in solution would become reduced in concentration with distance downstream from the source of pollution. It was also fairly obvious that the fallout of zinc as tailings that had been washed downstream and then deposited over the Foxlow plain would be, like the slimes dumps at Captains Flat itself, another intermittent source of high pollution from time to time, following heavy rains.

Commentary

The initial period of inspection, observation and experimentation of the Molonglo River was now over. It had taken several years and the major research time of at least four people, all of them scientifically qualified in the field of study. If the investigation and efforts had been initiated and managed by people lacking prior training and expertise, I think the problems of finding the appropriate approach would have been difficult if any clear cut understanding of what was happening in the Molonglo ecosystem were to emerge. For an amateur group to achieve a study of the type we did, they would have to have raised the funds to hire a professional consultancy at very considerable expense, or at the very least, to have worked very hard to acquire enough knowledge to organize and direct the efforts of the professionals they hired.

Subsequently

All we could now do was to report to The National Capital Development Commission what we had found and point out that pollution would continue, and that at least some zinc in variable concentrations would inevitably reach Lake Burley Griffin. We could not accurately predict that this would mean certain death to any fish in the lake. For that, much additional study would have been needed. We did strongly recommend that it would be most prudent to stabilize the mine wastes both in the slimes dumps at Captains Flat and at Foxlow. We noted that unless the amount of zinc from these sources could be much reduced, and their sites stabilized against the effects of future heavy rainfalls, it would likely, even over extended time, not greatly diminish the presence of zinc in the system. We noted that it had been found in England, in long-abandoned mine sites, that zinc-resistant grasses had covered mine waste deposits, thereby partly stabilizing them against erosion. Similar grasses were locally available near Canberra. The main tasks of stabilization we believed would be best left to mining engineers and to those equipped to mitigate the rise of mine waters from old, unstabilized sites, and who could advise on the restructuring of large surface waste deposits.

Unfortunately, that was as far as things progressed.

The National Capital Development Commission, who had funded part of this work, declared that with the completion of the elaborate dam that contained Lake Burley Griffin, plus the traffic bridges that crossed the lake and immediately became a vital part of Canberra's road system, funds for the lake's completion were now exhausted! This seemed a bizarre outcome. Why bother to fund and take an interest in the environmental future of the lake as being suitable for fish if the job was not going to be completed one way or the other?

We, who had worked so long on this project, could do nothing but express our disappointment, not only as scientists but as citizens of Canberra who knew better than anyone the conditions in the supply stream to the lake. The frustration we felt was the kind that many conservationists experience in trying to persuade responsible authorities to take action to safeguard ecosystem environments, even where the knowledge and means to do so are at hand. Appeals for help from government authorities with the necessary powers and duty under the law often go in vain. However, the last word on this subject had not been spoken.

Several years after we issued our report a newly elected Federal Government brought different attitudes to bear on a number of problems in Canberra. There was now some interest in returning to the question of what should be done about zinc in the lake, in which it had recently appeared that fairly regular fluctuations in its concentration were occurring.

Further information had showed up about 1971 in a survey of the benthic animals in the shallow water of the lake that was carried out by a number of students in an ecology course I directed. In this study only one species of mollusc was found and very few crustacean species. Odonata (dragonflies) and Ephemeroptera (mayflies) were also only rarely represented compared to what might be expected in an Australian inland water of this type.

By about 1970 I was looking again at the Molonglo and Lake Burley Griffin and considering many of the associated problems and, with Peta Dawson, a graduate student, published a report that reflected on these questions.

Matters arising

We now understood that zinc was gradually precipitated from the Molonglo river water through hydrolysis. However, we were now also aware of biogeochemical processes which ensured that zinc sulphate would be liberated from ores and mine wastes and would thus continue to enter the Molonglo. The slimes dumps at Captains Flat were composed of sulphide ore residues whose oxidation was linked to the activity of thionic bacteria which live in such deposits. *Thiobacillus ferroxidans* is capable of chemoautotrophic growth during oxidation of ferrous sulphate. And in deposits that contain ferrous sulphide this sulphate is continually formed. Through the action of *T. ferroxidans* ferric sulphide is oxidized to ferric sulphate, an aggressive solvent of chalcopyrite (copper sulphide) and sphalerite (zinc sulphide). By this means, requiring only water and atmospheric oxygen, the generation of zinc and/or copper sulphate and zinc sulphate is essentially constant, unless a method can be found to guard the sulphide from water and oxygen.

We had also become more aware that water temperature and toxicity of zinc are inversely related, meaning that fish would be less susceptible in winter.

The Queanbeyan River was popularly thought of as a tributary of the Molonglo, but in fact was the larger stream by a factor of 3. This meant that whatever the concentration of zinc ions in the Molonglo it would fall to one quarter of what it had been before it joined with the Queanbeyan. The zinc present in Lake Burley Griffin was apparently not at toxic levels. It was now thought that zinc in the lake would not exceed 0.4 ppm in a normal year. This presumably explained how the brown and rainbow trout, which were found in the lower reaches of the unpolluted Queanbeyan River, were also able to tolerate conditions in the lake. However, we were also aware of laboratory evidence that found trout unable to stand prolonged exposure to zinc exceeding 1 ppm. Given the other factors (water movement, hardness and temperature) noted above, and the possibility of occasional very heavy rains on the mine wastes at Captains Flat and Foxlow, we surmised that at times of maximum erosion of wastes, zinc levels intolerable to fish in the lake could be reached.

In 1973 my immediate interest in Lake Burley Griffin terminated, but a little earlier, Peta Dawson and I were predicting that, in the several years during the construction of a planned storage dam at Googong on the Queanbeyan River, when it was assumed the flow of its unpolluted water to the lake would be reduced or even halted, a significant rise in the zinc level in the lake might occur, which also might lead to trout mortalities.

In fact, however, there is apparently no record of a mortality peak of trout having occurred in the lake during this period. It is therefore reasonable to ask whether the water flow from the Queanbeyan into the lake was somehow maintained at a rate sufficient to keep zinc in the lake from reaching the acute toxicity level for fish. I have no information that answers this question, but I wish I had, because Peta Dawson, Les Penridge (another graduate student) and I examined the gills of 13 rainbow trout and 15 brown trout (all two to three years old) from Lake Burley Griffin and compared them with gills of brown trout from a hatchery in the State of Victoria. The Burley Griffin trout showed gill damage ranging from moderate to very severe. To say the least, the trout inhabiting Lake Burley Griffin could be assumed to have been significantly challenged in their normal activity by such damage, even though they were surviving. Such lesions as we found would not commonly be encountered in healthy fish populations. It does therefore appear likely that the trout in Lake Burley Griffin were not in ideal physical condition, at least at the time of our observations.

A little before this time, Peta Dawson and I put forwards a number of suggestions to eliminate or alleviate the continuing threat to the lake from its zinc-containing watershed. These were as follows.

1. A treatment plant could be installed in the Molonglo to remove zinc. The river could, for instance, be made to flow over crushed lime. However, this could be expensive and technically hard to maintain.

2. Chelating agents could be put in the river during or after floods to reduce the active zinc load. This had been known to work in Canadian rivers in reducing fish damage from copper and zinc.

3. The course of the upper Molonglo could be made to bypass the slimes dumps. This was technically feasible, but probably very expensive.

4. The dumps could be buried in the disused mine. Unfortunately much soil and rock fill had already been used to plug the mine shaft and its merits might, anyway, have been limited. It is one thing to have the zinc present in the rocks of the native ore body but quite another to return mine wastes to the ground in the unstructured and unconsolidated form in which it occurred in the slimes dumps. Zinc returned to the earth would probably have been even less stable than it was in the surface dumps. Water was still issuing from the ground at the mine site and during heavy rains could gush forth strongly.

5. Physical removal of the dumps to another location would have been possible, but probably opposed, as this would simply be making it someone else's problem.

6. It was asked whether the dump material could have any economic value. Could it be converted into a useful form, say for building? Nothing seemed promising, but the proposal had intrinsic merit.

7. An extensive, deep drain could have been excavated around the dumps to guard against release of runoff, and the dumps covered by a thin layer of some chemically inert material such as synthetic rubber. This suggestion was simple and direct in principle and would probably have been economically manageable. It could have required some of the steeper-sided dumps to be regraded to more stable surface slopes to minimize possibilities of collapse. But it also could have been a rapid solution and might have offered good control of the problem; see below.

Foxlow Plain would have required separate measures for its conservation. A possibility would have been to plough the affected area, liming and fertilizing the soil and planting zinc-resistant grass such as the *Agrostis tenuiis* known to be genetically resistant to heavy metal wastes in Britain, and to be locally available in the Canberra region. Such grass may form swards on mine-polluted soils, comparable in luxuriance to those on normal soils, if the soils are limed and fertilized. Another measure would have been to excavate shallow furrows parallel to the course of the river that would discharge to wider furrows that drained directly to the river. This would help remove rapidly major surface precipitation before it could soak heavily into, and destabilize the deposited material on Foxlow Plain.

As far as control measures at Captains Flat were concerned, all I know is that about 1974 it was realized that mine wastes had been swept 50 km downstream into Lake Burley Griffin itself, which resulted in a major reconstruction, regeneration and revegetation project in which dumps were reshaped, sludge pits were filled with clay and rocks and the whole was covered with grasses and other vegetation. This made a difference, of course, but the problem of heavy metals leaching into the river remains. These measures resembled some of those that Dawson and I had put forward in 1973. It should be noted that, in addition, zinc concentrations of about 2,000 ppm dry weight were found in the soils of many parts of the lake basin.

Not entirely anticipated

As for fish in Lake Burley Griffin, at least three native species have thrived over the last 30 or more years, including Australia's largest, most sought-after and legendary freshwater fish, the Murray cod, *Maccuullochella peelii*, plus five introduced species. I am not aware that anyone has critically examined the gills or other tissues of these fish, or those of the trout, in recent years.

In this long account I have tried to demonstrate in some detail and in a stepwise fashion how several people were able to reach a reasonably detailed understanding of the ecology of a zinc-polluted river. Our purpose was to gain enough ecological insight into this river's functioning to know its limitations and possibilities as a major contributor to the water supply of a lake in which it was hoped a recreational fishery could be established. I believe we were essentially successful in our efforts and that some of the suggestions we advanced for management and conservation of the lake into the future were realistic and workable in principle. Indeed, at least the basics of some of our main suggestions were applied. We were perhaps nearly as successful as might have been hoped for in suggesting ways to manage a complex ecological and conservation problem with limited resources of time, funds and investigators.

I also want to point out that any attempt at conservation—whether of a species or an ecosystem, and whether or not the species or ecosystem is completely natural or already modified or impacted by human activities—can never be relied on to come to a conclusion. Even as we thought our study of the Molonglo was more or less complete, new knowledge about the state of the system was accruing. Not only did trout survive in the lake, other fish of other species have also done so. And regardless of the measures that were eventually taken to restore and stabilize the zinc from Captains Flat, the Molonglo River, downstream to about Station 10, is reportedly still uninhabitable for trout. In fact many Canberra citizens recognize that Lake Burley Griffin is still not free from possible future catastrophe. It is, after all, intrinsically ridiculous to have a metal-polluted water supply entering a lake that divides Australia's Federal Capital city. In addition, there are reportedly moves afoot to enlarge the total size of the lake by an order of magnitude, even as others plan the grotesque measure of re-opening mining operations in the vicinity of Captains Flat! In such circumstances all the conservationist can do is be grateful that earlier investigations have at least furnished some real knowledge of the true state of the river's conditions. But with the uncertainty of future plans and actions that may unavoidably involve the Molonglo and the lake, all bets are off.

In closing, I must note that ecosystems tend to have blurry edges. What we recognize and name as an ecosystem is in the end a result of an essentially arbitrary act of judgement. Lake Dobson was a clear-cut example of an ecosystem, a basin of cold, clear water very low in dissolved solids, contained in dolerite. Until it was experimented on, it had little plankton and only very sparse vegetation of higher aquatic plants. Birds and mammals would have visited it to drink and some of their metabolic wastes would have ended up in it. Towards the other end of a spectrum of possibilities are ecosystems based on dense and productive forests visited periodically by birds that might have come to breed, feed or shelter from thousands of miles away in very different surroundings, only to depart again seasonally. Many large mammals wander over great distances and visit many ecosystems. Whales cruise the world's oceans, some of them moving between tropical and polar waters, feeding in various kinds of marine ecosystems. If, as is often done, the ecology of ecosystems is appraised in terms of the matter and energy that enters or leaves in the course of a year because of the activities of animals with roots or connections elsewhere, the true extent of any ecosystem can be very hard to define. I think that all I need to say is that anyone who is involved in understanding a particular ecosystem must have an appreciation of the degree of tenancy of all the creatures that spend time, though not necessarily their whole existence, in that ecosystem.

Successful ecological studies result from the manner of their planning. These days, much work on conservation is starved for funds and for want of professional ecological investigators to guide it. As a result, well-intentioned amateurs frequently find that, even if they can secure funding for conservation work, knowledge of how to conduct effective studies is lacking. The result can be massive wastage of precious money and failure from the scientific perspective. My aim here, and throughout this book, is to urge all conservationists to learn, or teach themselves, to plan the course of their investigations if the results are to be meaningful and worth their spending hard-earned research money. And to try very hard to get the advice and assistance of professional ecologists even if they are just retired professionals who will work merely from interest and devotion to the conservation cause—and not ask to be paid!

As conservationists we also need to embrace and garner outside interest in our activities. When I was very young some of the observations on mound ants were made immediately in front of my house. People in the house on the opposite side of the road saw me at work one day and I soon became aware of muttered comments.

"What's he doing?"

"Not sure. Counting ants or something."

"Is he sort of … ?"

"Maybe …"

I shrank, cast no glances and paid no heed—or tried not to.

Today, a lifetime of experience later, I would have smiled, waved, called "Good day," and beckoned my neighbours over to see what I was doing and hoping to share some of the interest and excitement I was feeling. Also hoping there might be some kids among them, because they are the ones you really want to make contact with. Because that's what you do. When people show interest you don't rebuff or ignore them, but try to get them to understand what you are doing and get their comments. That's how conservationists should behave if they want others to appreciate the world as they do.

CHAPTER 6

PAST, PRESENT AND FUTURE OF A LAKE AND RIVER

A Conservationist Perspective

The account given in this chapter departs from the earlier chapters in being the result of several studies performed in Canada. I participated as one of a number of people who, in recent years have tried to understand an already modified watershed system well enough to know how to conserve its qualities at least at their present standard. We have learned enough that, if events allow us, we can improve the system's prospects for survival as an environmental feature of great value and esteem for many people who know, use and cherish it.

Here is an attempt at an overview of a lake environment which is changing over time but which we hope to conserve as much as is feasible. Here are light, darkness, sun, stars, winds, rain, hail, snow, progression of the seasons, births, lifetimes, deaths, movement of many creatures—all under an open sky and daily turning from a beginning to an end and to another beginning. Not like a city where order is imposed, artificial, where the sky is fragmented by the lines and angles of non-living structures, time is an arrow, not a cycle. Here, in this river-lake system, you can think of space, time, plants and animals, Nature, eternity, not just now and then, but every day. This is Washademoak Lake which, to the great world, is like a grain of sand on a beach. So, can the conditions in Washademoak Lake and its watershed environment tell us things that can be applied to the whole of the world of Nature, or what remains of it in such times as troubled as the present?

Washademoak Lake, along with its much larger neighbour, Grand Lake, is part of the Saint John River drainage. This great river,

71

featuring in its lower half many complex waterways, including channels and sandy islands, extends from northeastern Maine, runs in New Brunswick, Canada, close to the U.S. border, traverses the cities of Fredericton and Saint John, and discharges into the Bay of Fundy. The connecting watersheds of the rivers tributary to the Saint John, including the discharges from the Canaan River and Washademoak Lake, are comprised of various landforms, environments and land practices, and are the collective homes of many species of plants and animals.

Washademoak is the second largest lake in New Brunswick. That does not mean it is very large as lakes go in North America. In width, it ranges from about five hundred metres to around 1 km for most of its 42 km length (except at Cambridge-Narrows Village where the opposing shores are no more than two hundred metres apart and near the foot where it is approximately 2.5 km wide). It could easily be mistaken for a wide river as it flows along slowly but easily.

The lake changes in colour with the light and weather, especially after heavy rain or wind. All the seasonal shifts in tree colour, from their most subtle to their most vivid, are on full display. You can't miss seeing the spring or fall leaves hereabouts. Birds and wildlife are here. The lake still has many species of fish. The things that are here may not be enough for everyone, but a basically self-sufficient person can breathe and see and ponder here in ways not usually experienced in cities. Can that be enough for you? Depends.

A detailed historical account of *People and Resources – Canaan-Washademoak Watershed* by Shawn Dalton and Robena Weatherley was published in 2005. It deals with an environment, which includes the Canaan River-Washademoak Lake watershed that was inhabited as long as six thousand years ago by ancestors of the indigenous Wolastoqiyik (or Maliseet) people.

The first Europeans (pre-Loyalist British) to come here settled on the rich alluvial meadows of the Canaan River (the largest stream tributary to the lake) in 1793 and were soon followed by United Empire Loyalists. These early settlers found the watershed rich in forests and in variety and numbers of game animals and fish. They were intending farmers but, to farm, their first efforts were directed to felling the great woods near the edge of the waterways to clear enough land for raising crops, mostly wheat, buckwheat and corn.

The cutting of the original Arboreal Forest began about two hundred years ago with giant pines fifty metres high, three and a half metres in diameter at the base, that were shipped to Britain as masts for naval vessels. As these great trees became depleted, trees of lesser size went to Britain as squared timber. Eventually, spruce trees less than fifteen centimetres in diameter were being cut and, for nearly a century, the New Brunswick forests have been harvested mainly for pulp production. The great resulting depletion of the forest was not only in terms of the permanent aesthetic loss of the towering woodlands, but also in the reduction of botanic diversity and habitat variety for bird and wildlife species.

As much as a century ago there was some awareness that indiscriminate cutting, driven only by immediate profits, would greatly damage the forest's ability to regenerate and to supply a variety of lumber to future markets, but this was largely ignored despite some government efforts at control. More recently, the attention of conservationists has turned again to maintaining biodiversity and the value of the forests to science, and increasingly to prospects for recreation, ecotourism and aesthetics. The public's right to a stronger voice in how forests are managed and conserved is gradually becoming acknowledged, though not without much opposition from those who continue to insist that cutting trees is the primary—if not almost the only—point of the existence of the forest.

By 1851, farms of Canaan-Washademoak were producing hay, wheat, oats, root vegetables, legumes, buckwheat, Indian corn, potatoes, butter, maple sugar, cattle, horses, swine and sheep. Grist mills and tanneries were established as were hand looms to make cloth. Even in the earliest days, animal protein was never scarce because of the abundant fish and game.

Family farm life was the keynote of existence on the watershed, established early in the 19th century and maintained well into the 20th century. Such farming led to an ample sufficiency of basic supplies, but for most families not much cash. Large-scale logging of the great forests and the establishment of functioning mills as early as 1814 helped alleviate the cash deficiency. Because of its strategic situation in the great Saint John River system the Canaan-Washademoak system offered excellent facilities for moving large-sized lumber south to the port city of Saint John.

With only a small amount of cultivated land near the water, farming families were still able to obtain drinking water from Washademoak in winter as recently as the 1930s.

It was only in the comparatively recent past that artificial chemical fertilizers were applied to offset depletion of natural soil nutrients. Volumes of silt have now been lost from increasingly aggressive forestry operations, and depletion of shoreline vegetation and heavy pressure of constant commercial fishing have produced severe and potentially permanent impacts on the quality of the watershed and the lake.

The chief threats to the quality of the water in the river and the lake that have occurred after WWII, following the economic collapse of the family farm, are mostly from soil disturbance resulting from the mechanically forceful methods of modern forest harvesting, including clear-cutting and other operations, together with forestry road construction, ditch digging and other forms of soil excavation. These activities can result in disastrous amounts of sediment entering the lake. And in the most recent years, many new arrivals to the lake have constructed buildings as retirement homes, recreational accommodation or as permanent dwellings from which they travel (often quite long distances) to and from their workplaces. Much silt has been added to the lake from the sites of many of these new buildings because of careless or unenlightened, sometimes wilful, disregard of proper building and land maintenance practices. All land misuse leads to degradation of the lake and its water quality and at present is the most threatening of the challenges to rational management and conservation of the watershed.

Because many people are nowadays seeking their physical recreation directly in the lake, it is not nearly as quiet as it used to be— especially on weekends and holidays—an added, and by no means minor, irritation to those who cherish peace and contemplation in their own leisure time. But the lake is still a splendid place to live near, regardless of the unpredictable levels of the spring freshets, when houses that are very close to the lake edge may have flooded basements and other damage.

Thirty years ago it was more pleasant to swim in the lake than now because progressive siltation has been providing a hospitable soil for rooted aquatic plants, and their shoreline fringe has greatly widened, making bathing more difficult and less inviting. Canoeing is now little practised because of the traffic of speedboats and jet skis. But there are still places where one can sit by the lake with the soft wind in the oaks and a clear blue sky above. And there are still soaring eagles, loons calling and the splash of an osprey hitting the lake surface, to come up spilling water, with a fish, that forcefully remind you that you are living in a world of still-surviving natural marvels.

Learning how to conserve Washademoak Lake

About 1989, local consciousnesses were raised, and some consciences were disturbed, when Washademoak Lake turned red following very heavy rains. This dramatic event excited—indeed outraged—lakeside dwellers and visitors who demanded an explanation from the New Brunswick Department of the Environment. Colour film of the lake, shot from a helicopter just after this rain, was screened at a large public meeting during which questions were raised, criticisms voiced and blame assigned. The atmosphere was fraught with anger and there was much misinformation. Even though it was clear that a great quantity of red silt had been eroded by the rain from the watershed's disturbed and fragile soils, and it was easy to see the main places it had come from, many in the crowd were questioning whether this was an enduring threat to the future of the lake and its watershed. The soil surfaces were especially destabilized where road-making and forestry practices had been clumsily conducted without due attention to appropriate "best management practices" (BMPs), and where there had been overcrowding and overgrazing of livestock on some tracts of agricultural land draining by a stream into the lake. There was also serious soil disturbance where land near watercourses had been cleared of vegetation and otherwise surface-damaged. Even now (in 2011)—twenty-two years later—it requires but a day or two of strong winds, not even needing the accompaniment of rain, to churn the water in this lake for the whole of its shallow upper half into becoming strongly red-hued again from the silt deposited years earlier. And significant rain at any time can cause the addition of more silt from places where drainage from exposed soil surfaces can reach the lake.

Though it was easy to point fingers and lay blame, the main challenge, clear to those who could be objective and thoughtful about what had happened, was to determine what should be done to ensure there would be no comparably-sized repetitions of the 1989 event. It was also clear that if this particular event pointed to certain large-scale examples of watershed mismanagement, it also served to draw attention to a multitude of smaller but comparable mismanagements whose collective effects over a period of years could have the same sort of total result as the initial catastrophe, if not worse.

The catastrophe aroused resentments, annoyance, frustration and anger in hundreds of people, many of whom did not even know each other. Many realized the lakeside ambiance—the wildlife, the woods, and the sparkle of this long stretch of water in which they bathed and boated—was one that in many other countries they would have needed great wealth to enjoy. So people were understandably fearful of the red water. Would it be permanent? Would it be a sign of further damage? Or would the lake revert to its norm? Whatever the case, there was, at least for a short time, a determination that something must be done, some plans laid for the lake's future—and very quickly.

Soon, the water did return much to its normal colour. Some small bays were left shallower than before silt deposition, and it could later be seen that shoreline aquatic vegetation had extended itself farther from the shore where there was newly deposited silt in which aquatic plants could take root. These things seemed superficial to many who had never seen the lake before the catastrophe. Yet the real warnings were clear enough. Unless action was taken to safeguard the integrity of the watershed's surface, silt would continue to reach the lake following rain, through the watercourses of its drainage system. And though coloration of the water might be transient, silt accumulation in the lake's basin, even if gradual, would be cumulative, and eventually disastrous.

In Europe, innumerable small lakes have been filled up with sediment over recent centuries, especially since the Industrial Revolution, as the pressures of growing populations have produced increases of all the activities that contribute to soil surface disturbance. It was with this sort of knowledge in mind that a few of those who continued to be truly concerned about the future of Washademoak Lake decided to take several steps. The first lake residents to initiate these steps were a small, newly established group, the Washademoak Environmentalists (WE). More recently this group has joined with the Canaan River Fish and Game Association and others as a founding member of the Canaan-Washademoak Watershed Association (CWWA). This has extended the focus to the rest of the watershed, the Canaan River and tributaries, and brought an increased membership of concerned citizens. The original plan of WE was to use modern technology, such as satellite imaging and GIS (geographic information systems, which involve aerial photography), plus direct, ground-level observation, to determine and record the physical features of the watershed of the lake, and perform chemical and biological analyses to assess the actual water quality over extended time. It would thus be possible to determine what was happening in the watershed landscape and how its features and surface changes related to the lake's water quality. Sufficient funds were raised to put together the outline of a plan in the hands of an environmental consultant who completed an effective study in 1997.

By this means, for the first time, a concrete picture of the main features of the watershed was obtained. Now it could be visualized how forestry practices, other land use, the distribution and density of roads and other constructions, increasing presence of lake shore dwellings, and a variety of other landscape components and structures bore upon the conditions in the lake and its water. The forested landscape was a mosaic which featured the effects of various forestry activities that had greatly reshaped its original form of two or more centuries ago. From this understanding something could be appreciated of the trends of future forest change in terms of composition and diversity of tree species, and of what the effects of this composition would be on the variety and abundance of birds and mammals, and of the forest cover's ability to stabilize surface soil against further erosion.

The advantage in having this information for the long-term conservation of the watershed was that it could facilitate future landscape management. It was, however (and still remains), a problem to find some way in which such information can be used to influence the management of those parts of the forested landscape that consist of privately owned woodlots of many different sizes that were, of course, and still are, subject to a variety of forestry practices. The study had been done essentially from the standpoint of "The Forest Ecosystem Design Process" (FED), the brainchild of a British environmental manager, Simon Bell. The advantage claimed for this approach in evaluating landscapes was that it gave ways to shape managed landscapes to fulfil different conceptions of their most desirable and attainable features and appearances. The consultant, in his 1997 report, wrote that: "The present forest structure has a broad range of species and age classes. This provides a variety of opportunities for wildlife habitat ..." However, he also noted that while "the general complexity of the area provides opportunities related to education, diversity and community involvement with a variety of stakeholders ... (the) complexity also makes it more difficult to learn about and understand the ecosystem."

Of course, if the large tracts of forested land and farmland around Washademoak Lake were under the control of one person or group, planning of such a design could be imposed and implemented fairly directly. With a mixed ownership, the major task becomes that of somehow arriving at a unifying image (for which it will be hoped to obtain general agreement) not only as to how the forested watershed should be managed to ensure the maintenance of good water quality, but also that its appearance should be pleasing.

However, since the early days of this study, many of those associated with it have essentially turned away from the idea of any large-scale application of forest ecosystem design. The FED may be useful in shaping, or re-shaping, a forest landscape that is already much affected by commercial logging, farming or other activities. But, as many now see it, the most desirable objective is to keep as much of the remaining wooded areas in a condition as close to natural as possible or to restore it as much as possible to that condition. Only by this means will enough of the watershed ecosystem ever be able to maintain species diversity among plants and animals, together with the most pleasing appearance of the forested landscape.

Water quality of Washademoak Lake

It was always believed that the "quality" of the water in the lake would be a ready indicator of the condition of the watershed. To this end, the entire watershed area was viewed as a number of sub-watersheds, their areas and shapes determined by the form of the drainage areas of the main streams that enter the lake around its periphery. The largest of these streams, by far, is the Canaan River, which enters the lake at its head, has its own sub-watersheds, and was itself the subject of a subsequent separate study. The data from water chemistry analyses helped identify the most unstable of the sub-watersheds.

The basic series of analyses, performed on water samples collected under ordinary conditions along the 42 km course of the lake, showed a tendency to become lower in levels of aluminum, phosphorus, colour and suspended solids with distance down-lake from the point of entry of the Canaan River. However, even in this simple series, it became clear that the chemical characteristics of some of the peripheral streams were causing the water quality in the body of the lake, adjacent to where they were discharging into it, to diverge appreciably from the main range of values. But it was in post-precipitation water samples (i.e. those collected after about 2 to 3 cm of rain) that certain quite vivid differences in quality of water inputs showed up, with values for aluminum, colour, suspended solids and turbidity becoming considerably elevated at the mouth of certain streams, acidity (pH) increasing in several, and in one stream both turbidity and phosphorus levels showing major increases.

The chemical data had thus shown, under both normal flow conditions, and those after heavy rain, that the water quality was influenced by differences among a dozen inflowing tributary streams. In the end, three streams were found to contribute disproportionately to lowering the quality of the water reaching the lake. Examination of the condition of the watershed by means of aerial photography and direct observations showed this was caused by runoff from disturbed surfaces of their sub-watersheds.

Fish of Washademoak Lake

If the condition of the watershed is reflected in its water quality, it is reasonable to consider whether the lake's organisms have undergone deterioration in diversity and abundance.

The fish of Washademoak Lake since about 1935 have numbered at least 22 species belonging to 15 families, some of them the object of commercial fishing. Ten of these species spawn in freshwater (anadromous), one (American eel) spawns in the sea (catadromous).

The remaining 11 species are either permanently resident in Washademoak Lake, or may spend part of their time in the estuarine reaches of the Saint John River or the sea. At the time of considering life history information of these 22 species, commercial fishermen were consulted; they reported that most of the species caught or observed had declined in abundance during the last two generations, which could have resulted from intensity of fishing or because of changes in the lake or the sea, depending on the life cycle of particular species. The most notable decline was that of striped bass, which were plentiful in the 1960s, but became scarce. Recently, striped bass populations in the Gulf of St. Lawrence are reported to have shown considerable ability to recover after over exploitation, but increased fishing in winter may intercept upstream movement of adult striped bass spawners to a significant extent. However, there was an apparent return of striped bass in Washademoak Lake during the 1990s. Eel numbers have reportedly declined, apparently in response to increased exploitation because of rising commercial value, though juvenile eels are apparently abundant. Catfish are also becoming more abundant. Speculation on reasons for these changes is pointless at present.

In attempting to surmise what the effects of changing environmental conditions might have had on fish by influencing their reproductive potential, we can consider temperature and food. Though there may have been slight warming of the lake and river in recent decades, it has been suggested that draw-down at Mactaquac Dam, which was constructed in 1967 and is located on the Saint John River about a hundred kilometres above Washademoak Lake, could flush out eggs of striped bass, which are semi-buoyant. Further, the low temperature of water released from the dam could reduce Saint John River temperatures by several degrees as far downstream as the point in the river where Washademoak Lake discharges into the river— enough to cause egg mortality. Five other species, Atlantic salmon, lake whitefish, rainbow smelt, common shiner and lake chub, have a limited southern range, suggesting they could have been affected by any significant local increase in temperature if that had occurred. This would apply most to Atlantic salmon and lake chub, whose southern limits to range are only just a little to the south of New Brunswick. But speculations about temperature's possible effects are unproven.

At least 13 species could be affected by changes in spawning areas or their eggs or hatched young being covered by ongoing siltation. Three species, which include the Atlantic salmon, could also have their eggs killed by silt deposited during the long period they must spend on the bottom during development. Two more species have adherent eggs that are deposited among plants where the effects of silt would be less. Though speculative, these remarks are based on facts known about the life histories of the various fish species, including especially their spawning requirements. At least they can indicate what factors may be operating on reproductive success for these species. And it is known that there has been extremely heavy silting of the lake in the recent past, and that silt tends to remain settled in the lake, except when some of it is temporarily resuspended by heavy wind, then redeposited. And, as already noted, additional quantities of silt frequently reach the lake from tributary streams and lake shore activities following rain.

It is particularly difficult to know how and whether changes in food supplies to fish may have affected their populations, but at least nine species of Washademoak fish markedly depend on zooplankton as food for their young. It is not known, however, how changes in the water quality of the lake may have affected zooplankton over the past fifty years.

All that should be said of the fish of Washademoak Lake, with the little specific knowledge at hand, is that though their overall numbers may be well below what they were in the1930s, the water quality changes and effects of siltation since then do not yet appear to have been severe enough to do fatal damage to the fish community.

The Canaan River

The next phase of the study focused on the Canaan River. It is by far the largest stream tributary to Washademoak Lake and its 148,000 ha watershed comprises 68 percent of the total Canaan River-Washademoak Lake watershed of 216,000 ha. The entire Canaan watershed can be divided into a total of 17 sub-watersheds that range in size from 2,200 ha to 26,000 ha.

Condition of water and watershed

The water quality characteristics of the Canaan River were first determined from a series of eight surveys made in 1997 in which water samples were collected from 11 stations along the length of the river. The condition of the land surface at each of the stations in the Canaan catchment was also noted. The main aim was to assess the likely major effects of water quality in this river on the future water quality of the lake of which it is the main supply stream. A part of the intention was to obtain baseline data on the Canaan River prior to anticipated likely changes in water quality resulting from projected highway and pipeline construction that would cross many streams tributary to the Canaan together with the river itself and associated forest.

An ecological land classification of the river/lake watershed shows it to be part of three ecoregions that differ to some extent in detail, but all of which feature underlying bedrock and soils that, when exposed, tend to be highly erodible and can lead to soil loss and stream sedimentation, particularly following heavy rain.

As with Washademoak Lake, much of the land use in the watershed of the Canaan has shifted from family farms and small woodlots to agribusiness and large-scale forestry. A particular hazard to the future of the entire river-lake system is the construction of many more permanent and seasonal residences very close to the water's edge that are too frequently built with partial, even total, disregard for conservation of riparian vegetation. The activities associated with building such structures therefore lead directly to discharge of sediments into the Canaan River and may reach Washademoak Lake.

In 1997, the forested area, in various stages of growth, maturity, clearance and regeneration, was 85 per cent of the total Canaan watershed surface. Wetland occupied 9 per cent and agriculture and dwellings were only 3 per cent.

The water was assayed for 31 chemical parameters and for fecal coliform bacteria and *E. coli*. The waters of this entire system are higher in aluminum and iron than those levels recommended by health authorities, but this is a function of the soil chemistry of the watershed and cannot be avoided or prevented. Otherwise, on a very broad view of the overall results, the water analyses showed only a few instances in which the established chemical or microbial maxima for acceptable water quality standards were exceeded. Chemical conditions in the Canaan were judged overall as not likely to be dangerous to aquatic life forms, apart from the silt lost from the soil after heavy rainfall.

On a few occasions, elevated amounts of coliform bacteria were found at some sampling stations after heavy rain, which were thought to have resulted from manure-spreading on agricultural land. Some elevated *E. coli* counts were also found, possibly due to leakage from an aerated sewage waste pond or faulty domestic sewerage systems. It was postulated that most such leakage would not have reached the river if the legally required riparian strip of riverside vegetation had been in place.

At this point in these studies, which were intended to form a basis for conservation, several best management practices (BMPs) could be suggested for the watershed:

(i)   More rigorous application of standard safeguards for road construction and drainage ditch clearing;

(ii) More care during site clearing in forestry practice, in conserving riverside and lakeside buffer zones, safer manure disposal, and proper design and installation of residential septic systems;

(iii) Application of BMPs should be a community-based responsibility to involve stakeholders so as to ensure widespread acceptance and adherence.

The water analyses for the entire watershed were continued for an additional three sets of samples in the Canaan River and Washademoak Lake, with the object of determining whether the quality conditions were remaining essentially constant in pattern, despite the presence of massive highway construction across the Canaan tributary streams which has now been completed. Chemical analyses of water from some 22 selected sites of the river and its tributaries, and along the body of the lake, have established that the chemical quality of the water, though subject to some seasonal variations for each sampling site, retains a characteristic seasonal pattern over time. An addition and an improvement to what was already being done would be to have provincial health inspectors check the system regularly for coliform and *E. coli* bacteria at selected "sensitive" spots.

A preliminary survey of the invertebrate organisms found in selected tributaries near their discharge into the lake was made in 2003. The small invertebrates that inhabit freshwaters include some types that are particularly susceptible to various forms of pollution. Thus, the presence of the aquatic stages of mayflies, caddis flies and stone flies are a very good indication that a stream or lake is not badly polluted. It was found these types of invertebrates were present in the streams monitored, giving further evidence that the system was, as yet, not greatly challenged. Also, the invertebrates in the Canaan did not differ much from those in some small streams in relatively pristine conditions in the watershed, that is, streams located above any possibility of serious pollution; this confirms the satisfactory biological conditions in the latter.

Fish of the Canaan River

During 2005 an electrofishing survey of 26 sites in tributaries of the Canaan River and Washademoak Lake was performed. Eleven species of fish were captured, of which seven have also been reported in Washademoak Lake, but four appear restricted to the Canaan. Of particular interest was the presence of young salmon (formerly thought to be very scarce in the Canaan system) at many sites, and of some brook trout. The total range of species in the Canaan-Washademoak system thus appears to be in the order of 27 species belonging to 15 families. This degree of variety of persisting species provides added assurance that the system is not yet too gravely affected by environmental changes to support a reasonably diverse fish fauna.

Of course, such evidence tells us nothing of the population dynamics of the various species which allow understanding of their abundance, productivity, size and age distribution, interactions between species and ability to maintain themselves over extended time. Such insights are vastly beyond the scope of the present work and would – as is always the case in obtaining detailed ecological understanding, require much time and great effort.

Recent and ongoing damage to the Canaan-Washademoak watershed

It was apparent to even the casual observer of the watershed, and particularly the lake shore, that the builders of many dwellings constructed on the lake shore in recent years had seriously, even completely, ignored the regulations that required the preservation of a legally-defined shoreline buffer zone of vegetation specifically intended to guard against sediment reaching the lake. Starting in 2003, a study was performed on the main stem of the Canaan-Washademoak system in which the entire impacted shoreline was inspected from a boat. Twenty-three of the more extreme examples were photographed and the exact locations of all other examples of serious shoreline damage were recorded. In all, 8.2 kilometres (7.2 per cent) of the total shoreline had been seriously damaged from the conservation point of view, and such an extent of damaged shore (which continues to increase) contributes very significantly to the lake's sedimentation problem. One has only to observe the shore after heavy rains to see tell-tale fans of reddish sediment at the lake edge, below damaged shore properties or below nearby roads or boat launching ramps. Subsequently, observations on shoreline damage were extended to some of the tributaries. Damage of a magnitude comparable to that found in the main stem was noted.

Certain creatures

The following notes did not result from formal investigations. They are intended to add to the sense of overall environmental qualities of the watershed, a very few of its intangible but priceless features.

When you live in or near wooded country you will not necessarily often see a lot of animals in a short space of time (though you may, if you live near a breeding territory for birds, a place where they are gathering to begin a southern migration or where there is a spawning ground for fish). Over an extended period, however, you do eventually see or come into contact with many kinds of animals.

In the surroundings of Washademoak Lake still live foxes, coyotes, bobcats, moose, deer, bears, raccoons, squirrels, ground squirrels, beavers, porcupines, skunks, groundhogs, muskrats, hares, moles, mice. During the course of a year about a hundred and forty bird species can be recorded near the lake, ospreys, bald eagles, owls, herons and crows prominent among them. Loons can still be heard and sometimes seen. That this sort of animal species diversity exists regardless of the amount of human settlement, cutting of the forests and the roaring of power boats on summer holidays can lead some to dispute that the lake and watershed are under any serious threat. But this misses the point, which is that conditions haven't yet deteriorated to a point beyond which cataclysmic change will become apparent. And, of course, a comparable array of creatures cannot usually be found in city and urban environments.

However, the continued existence of these and many other species, particularly of birds, depends on an adequate variety and supply of mixed woodland, including many hardwood species. Only by maintaining plentiful and diverse land can varied habitat conditions be maintained, including breeding refuge, shelter and food organisms (foliage, seeds, berries, insects and other invertebrates and small vertebrates). Predatory birds (i.e. ospreys, eagles, owls) living along shorelines must have high trees. Additional needs are protection from loud noise and other disturbing pursuits near the lake edge where loons, for example, are easily disturbed. The environmental needs for bears, deer and beavers are already too well known to warrant additional mention here. But constant watershed surveillance and safeguarding cannot be avoided if a rich assortment of wildlife is to be retained.

An outline of relevant demographics

The Canaan-Washademoak watershed is in transition, particularly in its lower reaches in the vicinity of the lake. It has already an elderly population (several years older than the national average) that will likely continue to increase in prominence, and will require appropriate support services and recreational and social opportunities. Many of these residents became members of the community between 1996 and 2001 signifying that this region was becoming a destination for retirees. Up to 39 per cent of the 1996 workforce in the area was employed in social services, education or a government agency, most of whom would have been commuting to other nearby communities to work. The limited time these people could spend at home would reduce their ability to participate in community initiatives and events. However, the population was increasing in this region. There are very few convenience stores in the lake watershed region and these could not support the full food and household needs of their communities. Though they do cater considerably to the requirements of passersby, trade at the stores is estimated to be about six times greater in summer than in winter. The region also lacks the banks, pharmacies and other institutions and amenities associated in larger communities with more constant customer demand. As populations increase a market may develop for many more major commodities than gasoline, liquor and other convenience store staples. It is also possible that some young people who grow up here will elect to stay, though at present the region offers little suitable economic or employment activity. Recently, the Village of Cambridge-Narrows updated its rural plan, and intends to become a location for economic growth based on aquatic activities and tourism over the next twenty-five years. The Chamber of Commerce is also intending to encourage a variety of cultural activities. With the recent acquisition by the Cambridge-Narrows Municipal Council of a magnificent piece of land in the heart of Cambridge-Narrows, which has now been declared a conservation park, the scope for cultural, educational and environmental activities should be greatly enhanced. Such shifts will call for Washademoak Lake to become a clean, safe amenity. It will therefore be critical for the community to protect and enhance its existing water quality and quantity. Whether to attract day-trippers, or those wishing to stay longer, the number and diversity of recreational opportunities, both on the lake and upstream in the Canaan River, will have to be increased. In any case, overnight accommodations will be required. There is already some disagreement between some local year-round residents and summer visitors concerning codes of conduct on land and water that are likely to increase in the continuing absence of some agreed-upon standards of conduct for the whole community or region.

The trends found in this watershed are not unique. The fate of the Canaan-Washademoak watershed has, since its first European settlement, mirrored that of New Brunswick as a whole in terms of its ecosystems and their human inhabitants. There is therefore much to be learned from this place: whether and how diverse groups of people can come together to preserve, protect and enhance the region they share, for the wide variety of opportunities that attracted them here, and which they presumably wish to continue to enjoy. But will enough of those who cherish the region be able to provide an organizational model for improving watershed stewardship and management?

What should be our conclusions and purposes? First, to bring to wider attention how it is possible to determine and record a lot of what is happening in a human-impacted watershed and, by extrapolation, in other impacted watersheds. And can we derive further knowledge of what can be done to safeguard against ever-continuing environmental degradation? Perhaps it can be argued that environmental conditions in this inhabited lake and river are "not so bad." But it can also still be seen how and why conditions could either worsen or be made to become considerably better. The Canaan-Washademoak waterway is a splendid natural asset. It received a powerful insult to its integrity about twenty years ago that was a wake-up call of what could go badly wrong. We are now in a position to prevent a repetition of anything resembling that disaster.

Our study of this system is ongoing and includes projects such as an "Atlas of Social and Economic Conditions in the Canaan-Washademoak Watershed," which will present a great deal of information on the current functioning of the area. Relevant information to residents has also been delivered in the form of a speakers' series. The CWWA has also conveyed information to schools throughout the watershed in the form of stream and forest field trips and classroom presentations. A video has been prepared that deals with watershed conditions. All of these endeavours are attempts to present information leading to better stewardship and conservation of our waterways.

Grains of sand

All the foregoing, not-very-technical account is about the history, development and present condition of a river-lake system, of which something of its recent past is known and in whose forested watershed there are various forms of habitation that house and shelter people and animals with a variety of reasons for being near the lake. We combine this mixture of facts with some recently-obtained information about how such events as flood rains and imperfect land use can produce deleterious effects on the quality and appearance of the lake's water and its potential as a continuing healthy habitat for humans and other organisms. Still lacking is a biologically detailed and comprehensive picture of the Canaan-Washademoak ecosystem, but it is no longer a mystery where anyone's guess about "what has been going on" is as good (or as poor!) as everyone else's. How much more can be learned will depend on how long, and at what cost and trouble, those who are really interested are willing to carry on.

It is now possible, over continuing years of monitoring and study, to understand the dynamics of process and change of this watershed well enough that the effects of most changes—good or bad—as results of new plans or activities can be foreseen and the magnitudes of their effects monitored and evaluated. The results can then be added to and built back into our ever-increasing knowledge of the system. Mere knowledge will not, of course, guarantee wise stewardship. Take mining as an example. In this technological age there can be greatly detailed knowledge of the feasibility and planning of a mining operation well in advance of its actual existence. But in the long run, its wealth and productivity, its impact on the adjacent landscape and on the lives and welfare of its workers and the non-mining inhabitants who live nearby will become matters of human stewardship. This stewardship will call, above all, for good judgement, ethics and high civic responsibility.

In the end, perhaps there can actually be use in a "grains of sand" metaphor. These days often seem like conservation's last moment to mount a global defence against forces bent on destruction of the world that we have known, forces that are fearlessly assuming to replace it with exactly "what" has never been globally discussed or agreed upon. But whatever it is will be both alien to past experience, and it will be final! If ecosystems or watersheds (which may only be parts of ecosystems) can be thought of as entities that we can rationally investigate and eventually achieve real stewardship over, they are, on a global scale, as grains of sand. By this we mean that they are different from each other, yet with points of similarity. Every time we learn to manage and conserve one watershed, it will equip us with some of the concepts and skills required to conserve many others, no matter how large, remote and mysterious they may seem. We should not seek to homogenize them. Grains of sand are rock fragments that can seem alike, but under the microscope are seen as unique, marked, shaped and sized by their origins and their histories. Yet, if geologists or chemists were to study individual sand grains, they would inevitably use similar investigative methods on them all. The elements of which they are composed may be put together in a thousand different ways. But to study a few will help us to understand why others have different colours, chemical compositions, hardnesses, are of varying age, have even come from different historical exposures to the forces of Nature. It is the same for ecosystems or watersheds.

A critical comment

One thing has constantly disturbed us. We (the CWWA) are very grateful for much help and advice rendered by scientific and technical staff of the New Brunswick Department of the Environment. But we regret that no more than minimal action has been taken in recent years against persons who ignore, even in a most egregious fashion, the existing regulations to protect the edges of the lake or river as supposedly enforceable by law. In addition, there appears to be little or no difficulty in obtaining permits that allow people to construct buildings or alter waterways and shorelines in ways that should under no circumstances be allowed. This is extremely unfortunate because, though some people who break environmental laws are sorry when their offences are pointed out to them, and wish to make amends, others may be contemptuous of the law and of the opinions of their neighbours. This is creating a problem that must be addressed! But it is something that unpaid volunteers in environmental stewardship lack the authority to act on. Only employees of government departments have the legal status to deal with these matters. Regrettably, all too often they appear to find it impossible to apply the law even in extreme examples of environmental abuse. This must change!

CHAPTER 7

WHAT HAPPENED TO THE GARDEN?

We are billion-year-old carbon
And we've got to get ourselves
Back to the garden.

Joni Mitchell

The dilemma of the conservationists

I want to move now from ecological and conservation problems of limited size and local concern to consider some of more global significance. These problems are too massive for most voluntary conservation organizations to work on and really need the efforts of national and international organizations with the money, skilled personnel and the mandates to tackle them. Despite their magnitude, they are still problems of ecosystem conservation and can be understood in the same manner as the local problems of limited scope that concern most of us most of the time.

Either from their own experience or from media reports, millions of people know that very bad things have been happening to the planet. They know that the survival of many of the creatures that inhabit the world's environments are under threat. They know that conservation is working neither well enough nor fast enough. Yet there does seem to be large public good will towards aspects of conservation, but it's all too often focused on particular species of spectacular, dangerous, noble, rare, beautiful or unusual plants and animals and especially "warm and cuddly" or "cute" birds and mammals. In addition, the good will usually fades under warnings (many of them false) that conservation measures as proposed, for

91

example, for a forest landscape, may threaten people with loss of jobs, income or welfare. Governments could, of course, assure protection and compensation for those whose jobs or homes were under threat from a particular act of conservation, but most governments decline to offer such assurance. As a result, businesses that are concerned mainly with profits are provided with constant leverage against any type of conservation that they do not want.

A host of environmentalists with high credentials are straining to promulgate the urgency of conserving Nature. But as conservationists admit, even many of the most fervent, closely-argued appeals are failing to create the broad popular and political base needed to influence events in their favour, as Shellenberger and Nordhausin have noted in their essay *The Death of Environmentalism*. It's not that environmental and conservation data are deficient or inaccurate: the recent Millennial Environmental Assessment (MEA), the result of the efforts of 1,300 ecologists, makes that clear. But on the vast scale that is called for things are not going as they must in order to be widely and genuinely effective in saving the natural life forms of this planet. New approaches must be devised without delay.

As for an increase in the power or capability of government institutions or agencies to decide and act positively on matters of conservation, we cannot be confident. Such agencies, which might be assumed to be mandated by their very existence to support and even initiate conservation, seem more than likely to ignore, or anyway fail to react to, evidence and sound advice. It appears that this failure to act is largely because of pressures exerted by their political masters and often, on them, by business and industry. Let me qualify what I said a little earlier: sometimes conservation measures are brought about through the efforts of determined NGOs when they can somehow exercise enough logic, moral persuasion and even ethics to get government agencies—or even politicians who control them—to do the right thing. At present, however, getting enough volunteers to help change the way the world should go is a daunting challenge. It calls for far stronger and more charismatic leadership than is common even among devoted conservationists.

The magnitude of the situation conservationists are facing has been recently expressed by the World Conservation Union, the International Biological Congress and thousands of individual biologists: up to 20 per cent of all living species are predicted to disappear within 30 years due to resource extraction, industrial agriculture, introduction of exotic species, human population increase and other causes. This has been named a global mass extinction. The main difficulty in deriving benefit from such statistics is this: 70 per cent of the membership of the American Institute of Biological Sciences believes in the likelihood of this order of extinction, but a Harris Poll found that 60 per cent of the public were still unaware of an impending biological collapse. As the recent MEA has stated, "Over the past few hundred years, humans have increased the species extinction rate by as much as 1,000 times background rates typical over the planet's history." Some believe that, for effective global results, it will be necessary to wait until poor countries approach developed countries in general economic affluence. This would be a tragic blunder, for while presently the rich countries may do the most environmental damage, poorer countries, as they strive to match the industrial output of the rich, will add, by an order of magnitude, to the total damage done. As China, India, and perhaps South America and parts of Africa attempt to come "on line" as major industrial powers the total effect of this, driven by their already huge population numbers, will simply be to hasten intolerable planetary damage.

While they try to find outstandingly persuasive, charismatic communicators for their cause, present conservationists must continue trying to obtain whatever interest and help they can from those politicians, business leaders and industrialists who happen to possess the sustained interest and knowledge to pursue and augment conservation ideals and goals. That this can occasionally be done is exemplified in the recent rare achievement of saving a patch of rain forest on the coast of British Columbia from logging and other incursions, thus safeguarding mature forest, its contained wildlife fauna, including the magnificent and non-plentiful "spirit bear," from all hunting. This was managed by a seemingly unlikely accord among logging interests, native people of the region, and conservationists. So significant achievements are sometimes possible, even when the odds against their success may seem formidable.

But, in general, the message for conservationists is that they must somehow infuse a large public with the sort of "reverence for life" that Albert Schweitzer used to call for. This does not mean that only people who match Schweitzer's singularly uniquely charismatic appeal can effectively deliver such a message. In fact, it is very difficult to state exactly what the qualifications and qualities of the new conservation leaders must be. But without in any way belittling the heroic and selfless efforts of so many respected current leaders of conservation, we must find communicators who—while possessing voluminous knowledge and awareness of environmental matters, and the keenest instinct of how and when to use such knowledge—will have an extra dimension to their personalities. They will, of course, need great skill in public debate, but also the sincerity, warmth and passion to touch people. The efforts of present conservationists can never be underrated, but to obtain the greatest results from their efforts persons are needed who are charismatic leaders—persons with great communication and leadership skills who can carry the conservation message to the masses and the masses with them. Such advocates will assume the mission of inspiring a great societal and eventually global response. We have already people such as Al Gore, David Suzuki and Elizabeth May. But let us now begin "desperately seeking" many more of these people. For when the avid support and enduring demand of a broad public exists, politicians themselves will assume the roles of conservationists and environmentalists. And then conservation will become "merely" a major part of the mandates of all governments.

However, no matter how effective the leaders of the environmental cause might become in delivering their message, it will remain vital that the work of all serious conservationists bear reference to a manifesto simple enough for anyone to understand and be seriously idealistic enough to be universally respected. The manifesto must express reasons for conservation of Nature that will be fully understandable and will not contain any hidden, cabal-driven agendas known only among leaders in the field. The thrust must not be to save a world for a minority with elitist insights about environmental conditions, but to engage a great public who will know that the reasons to respect, cherish and conserve what remains of the natural wildernesses and life forms on this Earth are because they constitute the one true home the human family came from. This must never be construed as a sort of reshaping to resemble some version of a simpler and more agrarian past. In fact, given the present world population and the great interactive ecological subtleties that global conservation will demand, a conserved world would be more scientifically and technically demanding than the present one. It would certainly not be a world in which city and urban living would become neglected, but rather enlightened.

The problem is how to bring people to see why environmentalism and conservation, which critics try to ridicule and discredit, are subjects in which conservationists should strive to engage everyone.

Revisiting Nature

The reason conservation is not taken really seriously by more people is because it is still not generally appreciated that, for most of our existence as a species, we truly lived as a part of wild Nature, amidst a dazzling variety of plants and animals that evolved and survived with us over the millennia in response to the world's various topographies and climates.

Conservationists need to tap into humans' long-neglected feelings for the objects, organisms and processes of Nature. It is actually difficult to understand why great neglect of Nature has occurred for so long. For the fact is that, even if it might not always seem immediately evident in this highly technological world, a great many people do respond with positive emotion in the presence of giant old-growth forests, mountains, unspoiled rivers, the space and stillness of great canyons or rocky sea-splashed shores, even deserts or icy wastes. And, of course, they also respond to wonderful plants and animals with feelings that range from simple awe to what feels like recognition of fond memories. And everywhere, we can see perpetually huge sales of Nature images on postcards, in diaries, calendars, coffee-table books. And these images are stirring even to many city-dwellers who may never even have seen their actual subjects. This is surely because our brains are hard-wired to respond to them. They are the images of various versions of our ancient environments, the places where our species lived and travelled, hunted, found food and refuge as we evolved and wandered for periods of great duration.

Amita Sinhar of the University of Illinois, an environmental psychologist commenting on studies concerned with landscape perception and appreciation, has noted that images of wild Nature speak to people with directness and poignancy. This stimulation can occur even among many who have become habituated to landscapes of a modified, tamed Nature, but who are still seemingly genetically determined to like landscapes of natural wildernesses. John D. Barrow ("Artful Universe") has come to similar conclusions in mentioning human preferences for parks and gardens that tend to resemble African lands where much of the childhood of our species was passed. These are places characterized by broad sight-lines of the kind that enabled our ancestors to see both enemies and prey, while affording them the shelter of vegetation patches that included trees. Barrow also notes that high places offered locations of shelter and defence, which accounts for our pleasure at seeing rocks, hills and mountains—and even abandoned massive buildings, such as castles and forts.

Russian artists Vitaly Komar and Alexander Melamid determined in opinion polls what people liked and disliked in the visual arts. Among Americans, landscape pictures were the overwhelming favourite, with preference for water, forest and blue mountains, and in the foreground a single tree, deer and a human family. Polls in Europe, Britain, Australia, Africa and Asia found similar preferences, with local variations. Melamid and Komar also noted that this kind of picture is "the visual common denominator of people's imagination ... a picture of the communality of (their) minds ..."

It seems a reasonable inference then that contemporary impressions of natural landscapes derive from a universal consciousness and are therefore pleasing, beautiful or wonderful to a point where even those who do not in any way consider themselves to be conservationists would greatly regret their disappearance. It is in this sense that many people may be unconscious conservationists and why many are removing themselves from cities to live in the countryside, even if to do so means commuting long distances and may eventually drain cities of much of their vital raison d'être.

Our period as hunter-gatherers living in natural ecosystems lasted for many thousands of years. But though there are still a few examples remaining in the present world, when the period as a whole came to an end we had essentially completed our biological evolution. This means that we emerged as efficient long-distance walkers, diurnal creatures with good binocular vision and hearing, and hands equipped with opposable thumbs that provided us with remarkable dexterity and gripping power. Our highly developed but unspecialized hands enabled humans to lift, carry and manipulate objects of a wide variety of shapes, weights and sizes. And because our hunter-gatherer species was neither small nor weak, and hunted in family or tribal units, our ancestors could sometimes use simple weapons to kill large animals, which provided quantities of meat that their teeth and alimentary systems had evolved to deal with and which increased the variety and richness of their diet.

We are, of course, natural omnivores as shown by our teeth, acid stomach, digestive enzymes and by the overall morphology of our alimentary organs. But the hunter-gatherer life involves chance and opportunity, and its practitioners may have to change their locations seasonally or as local food supplies become exhausted. How would such demands have affected human welfare?

One might suppose that the lives of hunter-gatherers would involve a constant struggle to obtain food, but arguments and evidence from Jared Diamond of UCLA and from William R. Leonard, an anthropologist from Northwestern University, indicate that humans, forced to live on animals at those times when a more "balanced" diet containing more vegetable food was unobtainable, may have been at least as calorically well-nourished as most of the present human population. They may also have been healthier in terms of their body mass indices and blood cholesterol levels. Much of Leonard's evidence comes from the known diets of the remaining few hunter populations in Africa, North America and Russia (where the Russian Evenki follow herds of semi-wild reindeer). Recent studies have also shown that the Inuit can survive in health while obtaining more than half their food calories from wild animals, the fats of which tend to be in less saturated (often monosaturated or polyunsaturated) forms than those in the diets of modern North Americans, and others from countries of supposedly high living standards.

During the long wanderings and migrations of our species (vividly described by Jared Diamond in *Guns, Germs and Steel*), our ancestors came on a great range of topographies and climates. Living conditions will have ranged between the extremes of "Garden of Eden-like" biological variety and plenty to the harsh conditions of deserts and polar regions. One might assume that "Eden-like" conditions would have been the most attractive overall, conditions such as those of the Amazon basin or ocean beaches backed by richly populated forests. And certainly, favourable impressions of such environments seem deeply etched in the imagery of literature and art. But from the cultures of people in such extreme environments as deserts and polar regions, it appears that they, too, hold feelings of loyalty and affection for wherever they have been able to count on the continuing security of shelter, food and familiarity of surroundings.

What turned us away from Nature?

Jared Diamond also explains that farming, our first complex technology, had many diverse origins and various courses of development in different parts of the world. Thus its beginnings occurred over a great range of historical times, sometimes following a gradual introduction while people still partly retained a hunter-gatherer lifestyle, at other times temporarily abandoned that style, then reverted to it, then abandoned it again and so on. Sometimes farming has been smoothly transferred from one culture to another, at other times it has had to be completely rediscovered by particular cultures.

The world history of farming has also been profoundly influenced by differences in climate, soil, availability of cultivable food plants and animals suitable for domestication, the circumstances of the origin and manufacture of appropriate tools and the degree of isolation experienced by some cultures. As the first villages and towns multiplied many natural landscapes may have begun to seem inimical to agriculture, eventually setting up a habit of taking Nature for granted, especially as there might seem to be plenty of it left for future exploration or use. On this view of events, Nature, in its more or less wild form, would gradually come to seem a thing of the past. Then human society and its close-quartered life in towns, villages and eventually cities would be thespian and circumference of significant human activities, agriculture itself remaining as vital to existence but no longer the central drive of urban life.

But Nature still lives on in us

Let us, however, return to the effects of wild Nature on our ancient mind-sets.

Consider forests.

About ten thousand years ago much of the Earth's surface was forested. About half of that original forest remains and only about half of that is in its approximately original form, as mature Boreal softwood forests of Canada and Russia. There are species-rich tropical hardwood forests in the Amazon region, and there are other, but much smaller, patches of original forest in the Far East, Australia, New Guinea, Japan, Africa and in various highlands in which commercial exploitation is, so far, impossible.

Highly populated Europe has had fragments of its once great forests surviving into recent times, still with human inhabitants, demonstrating that forests can still be lived in in something resembling a state of Nature. Simon Schama has related in *Landscape and Memory* how the vast primal forest of Biatowieza in Lithuania, formerly the homeland of abundant bird and animal life, including the once numerous European bison, was fought over since the late Middle Ages by Lithuanians and Poles, Russians, and again by Russians and Germans during the 20th century. Schama recounts how this great natural woodland was fiercely plundered for trees and mammals, except for a few serious attempts at forest and wildlife management and conservation. Yet, as late as the early 19th century, there were forest people (who) with their nut-brown weather-beaten faces … conspicuously disdained the drudgery of the fields for (a life of) hunting and gathering, much the same as their pagan Lithuanian ancestors. Their dwellings, sometimes deep within the woods, were log cabins (and their) knowledge of the ancient forest was so intimate and so intricate it allowed them to subsist handsomely on the most succulent wild mushrooms, on the intensely fragrant tiny bog cranberries, aromatic wild woodland honey, broad leaves of sorrel and bulbs of wild garlic.

These people were foresters, gamekeepers and beaters attached to the Royal Hunt. In return for a paltry sum paid each year to the government, they were allowed to take any game they wanted within their district (excepting elk and bison), plus pelts of otter, badger, ermine, beaver and marten. Pelts fetched prices that paid for their licences and supplies of vodka.

Thus, though the life lived by these latter-day forest dwellers was not strictly that of naive hunter-gatherers, they were able, with the aid of very few outside connections, to enjoy an abundant material existence with plentiful food and shelter while relying almost exclusively on the bounty of the great forest. This picture manages to convey a more immediately appealing impression of the efficacy and possibilities of a hunter-gatherer existence than those of the few peoples, such as some Inuit, Amazonian natives or Australian aborigines, who still practice such a life style today, but whose ways we may have difficulty in comprehending. For their ancestors long ago settled ecologically, evolutionarily and historically into forest environments, whereas the recent dwellers in the Lithuania forest were people who, though in some contact with the developing world, were able to hang on in the forest with benefits. Thanks to Schama, I think we can form a much clearer idea of how forest life could be an abundant one without straining our imagination.

Schama further relates how, in 1943, Jewish escapees from ghettos reached the woodlands to establish a "primitive community of equals, living in pits covered with branches and moss or abandoned woodsmen's huts," again demonstrating that even modern people were able to subsist in parts of the ancient forests that had not been so degraded that most of their basic resources had been destroyed.

Facts like this build towards powerful national attitudes and mythologies. Two thousand years ago the Germans successfully defended themselves against the Romans for two centuries in their great forests. This knowledge lies at least partly at the heart of the enduring attitude of Germans towards their forests to this day, even though those forests are nowadays mostly the results of advanced silviculture.

Schama notes, on the other hand, that neither the Greeks nor Romans in the heyday of their civilizations appear to have greatly revered the primitive forest, preferring the "Arcadian" ideal of the variously settled and wooded countryside. However, both the Greeks and Romans had certain earlier traditions that did allot a significant place to the natural forest. Schama also describes the "greenwood," or comparatively open and modified remains of the natural forest in the history of England. Here, too, the woods were of great national significance as game and hunting reserves, as hiding and dwelling places for those seeking refuge, and as sources of oak for the British navy's ships. Wood was also a source of revenue both for various British monarchs, beginning with the Tudors, and for many other landholders who made fortunes based on wood supplies. Perhaps the British attitude towards woodlands (most of the great natural British forests were decimated by the Middle Ages) emphasized their roles as hunting reserves, beginning with the Norman Conquest and their economic worth. The British have a love of trees—especially oaks, because of their former industrial employment, but probably never, in the last two millennia, with the force and spirituality of the Europeans to whom the very impenetrability of the primal woods long meant shelter, home ... and refuge.

And then the depredations of recent ages

Near the end of *The Great Gatsby*, F. Scott Fitzgerald wrote meditatively of Long Island as first sighted by Europeans:

> "And as the moon rose higher the inessential houses began to melt .away until gradually I became aware of the old island here that flowered once for Dutch sailors' eyes—a fresh green breast of the new world. Its vanished trees, the trees that had made way for Gatsby's house, had once pandered in whispers to the last and greatest of all human dreams: for a transitory enchanted moment man must have held his breath in the presence of this continent, compelled into an aesthetic contemplation he neither understood nor desired, face to face for the last time in history with something commensurate to his capacity for wonder."

Actually, a number of historians and artists following the times referred to by Fitzgerald were very conscious of the magnitude and majestic beauty of the American forests which were celebrated in poetry and literature, prayed over, painted and speculated about. But many more Americans were quick to put the axe to them. This, of course, was partly unavoidable—the society had to have wood—though it is hard to banish cynicism when, as Schama notes, many of the choicest forest landscape paintings were acquired by the very timber barons whose brutal assaults were already making mayhem among the trees that were the subjects of the paintings. Later, the sequoias and redwoods of the Pacific coastal forests were also seen as vegetation of a magnitude new to Europeans and it must have required a major insensitivity among at least some woodland operators to view these giants as just more trees for the felling. So great were the Western trees' ability to generate wonder that they attracted many people to travel long distances to see, touch and make pictures of them. In 1864, in the midst of the Civil War, President Abraham Lincoln signed "an unprecedented bill" that granted "Yosemite Valley and the Big Trees" to the State of California "for the benefit of the people, for their resort and recreation, to hold them inalienable for all time." But if popular sentiment was properly impressed by the notion of the transcendent, perhaps sacred, qualities of such never-before-seen vegetation, other, less-refined sensibilities simply accepted the economic riches they represented and were rapidly plundered for. In the heartland, by 1890, trees were almost gone from Wisconsin, and Americans were buying rights in Canada, according to an article by Peter Unwin writing in *The Beaver* (*A Brief History of Trees*). And, by 1920, "the spoliation of the Lake Superior forests was complete." The aggressively mercantile and entrepreneurial types, who have always figured so prominently in the settlement of America, overwhelmed the contemplative aesthetes, who might have hoped to exercise some powerful conservation impulses among the early settlers. According to Rosenberg and Birdzell, writing in *The Scientific American* in 1990: "Europeans visiting the U.S. in the mid-19th century often criticized Americans for wasting natural resources. U.S. agricultural techniques did frequently lead to rapid declines in soil fertility. But more land was always available, so the losses were supportable. Americans invented wood-working machinery that appeared extremely wasteful to the British. At the time, however, it made good economic sense in a country so richly endowed with forests." This statement is as dismal as it is misconceived.

The Canadians matched the attack by the Americans. Peter Unwin writes that through the work of the timber barons in the period of 1870 to 1910, the largest pinery on the planet was cut down and shipped east on a hundred different rivers. It was an act that staggers the imagination, cost the lives of an untold number of men, and was perhaps as grandiose and labour-intensive as the building of the great pyramids of Egypt or putting a space ship on the moon.

Much of the early cutting was in forests of gigantic trees. Unwin mentions an oak in southern Ontario, with branches that started twenty-four meters from the ground, and which may have been left standing only because, at the time, Canada had no saws long enough to cut it.

The Canadian attitude towards forests today is somewhat conflicted. Trees are so much a national symbol that Canada's flag carries a red maple leaf. Yet huge ongoing arguments continue between those who demand the complete conservation of some old growth forest that still remain and those at the other extreme who believe nothing should be protected from commercial exploitation merely because of its antiquity, or its scientific and aesthetic interest. If experiences have failed to elicit a surge of anger at the latter attitude, they should at least have produced a groan of disappointment that the painful environmental misdeeds and bitter experience of the Europeans failed to result in better stewardship among those of their ilk who would inhabit the New World. (The recent experience of conservation of a large area of coastal forest in British Columbia has already been mentioned as a worthy exception.)

But even if we can demonstrate that there may be much (buried) love of Nature among the public and also understand—using mainly the case of damaged forests—that the time available for serious intervention is becoming increasingly limited, the question remains: what will generate the necessary public will and determination to save what remains of Nature on a sufficiently massive scale to avoid its almost total loss and ruin?

Rediscovering Nature

I believe that a conscious affinity with Nature forms the shield of Perseus through which man can affront the Gorgon of his fate . . .

Cyril Connolly (The Unquiet Grave)

A few centuries ago, at a time that featured significant advances in hull and sail design in ships and in reliable navigational aids, with the resultant opening up of oceanic trade routes, an heroic age of global exploration began. Mariners, emboldened by their better equipment, ventured farther from their home shores to make systematic voyages of discovery and produced maps of unprecedented detail and accuracy. This period culminated in the great discovery voyages of the 18th century, when scientific mariners, like James Cook, with scientist companions, explored the Pacific. Once knowing they could get around the world with some reliability, explorer-naturalists of the 19th century began the scientific study of wild nature in such places as Australia, New Zealand, the Pacific islands, South East Asia, Africa and South America. And of all of them it was probably Charles Darwin and Alfred Russel Wallace who eventually stimulated humans to come face to face again with Nature.

And what did they experience when, together with a host of lesser, but still curious and wondering, wanderers, they began to see the world whole for the first time in history? Well, regardless of their differences in nationality, language and basic knowledge of science, what they must have experienced collectively—whether or not they all grasped it—was being able to view an Earth in which wildernesses and wild abundances and great natural variety were still in advance of human depredations. They crossed the North American prairies and saw buffalo herds and pronghorn antelope. They travelled the plains of Africa with their teeming, miraculous wildlife and entered the great forests with chimpanzees and gorillas. They marvelled at the biodiversity of coral reefs; saw the pampas and soaring mountains of South America and the vast Amazon basin with its incomparable variety of plant and animal species. On the dry island continent of Australia, with its fierce blue skies and its unique trees they found marsupial mammals that were divergent in evolutionary origin, but still resembled, in many ways, forms and habits of the eutherian mammals found in most other parts of the world: these then were matchless examples of parallel evolution among vertebrates. And everywhere, the explorers would see many of the world's native humans, living in the great variety of conditions they had experienced for millennia, sometimes in comparative comfort and ease, sometimes in desperate poverty; some still as their prehistoric ancestors had been, lacking luxuries, but often with mythologies and deep concepts about space, time and the other creatures of their environments.

Since those times of the first great world bio-exploration key concepts have been developed that provide a formal structure to the science of ecology. These concepts include population dynamics, community, succession, climax and ecosystem. The meanings of these terms are not at all obscure, yet have required an unaccountably long time to establish themselves in the minds of many otherwise educated people. Largely as a result of this time lag, Nature is still commonly conceived as wild, disordered and chaotic, though in fact it is subject to principles of patterning and order dictated by physical conditions, by the relationships and interactions between species, and between and within populations, and also by the seasons and by the physical and chemical conditions of the environment. To understand something of these relationships and interactions is to understand how ecosystems function.

Our early ancestors inhabited the natural landscapes found within (or as parts of) ecosystems, surroundings where they might pass their entire existence, unless conditions changed so as to threaten them. They knew the edible plants and animals in these places. They knew nothing of the microbes science has discovered, or the chemistry of photosynthesis, or very much about agriculture. But in a human lifetime they would have acquired enough understanding of their environments and how to stay alive in them in a certain state of harmony, or at least of balance. They would have learned to associate a food plant's occurrence with certain types of soil, shade, the effects on it of wind, water and sunlight, and something of its relationships with other plants. Jared Diamond, based on first-hand experience of New Guineans, claims that primitive hunter-gatherers would easily have learned to identify scores, if not hundreds, of edible wild plant species in their environments. Our ancestors would have learned the habits and whereabouts of animals they could kill and eat or those they needed to avoid. The more thoughtful among them, like those among the Australian aborigines before the British arrived, would have contemplated the interconnectedness of the conditions in the ecosystems in which they lived. They would have noted that rain which fell on upland slopes led to growth in previously parched areas where animals they hunted for food (birds, kangaroos) congregated. They could understand that streams made their way to the sea, that they would need to adjust their lifestyles and locations to match different seasons, or sometimes to long-term shifts in the weather which could also lead to major changes in the distribution and abundance of animals and plants they consumed. At the more difficult times, their survival would depend on experience, or decisions to move on.

Some will dispute that our forebears lived in harmony with Nature, claiming that all animals including humans exploit their environment and that any species will tend to increase its numbers, expand its range and hence increase its environmental impact. On this view, any so-called harmony between us and other organisms in Nature will merely be the condition when food is plentiful and our numbers are low enough to avoid overcrowding. With this view, the present rampaging increase in human numbers is therefore "normal" enough, being simply a sign that we are using our space and supplies of food and other commodities to the limit of our technological expertise. But even if such a view can partly account for our situation in the world, it is nevertheless intolerable, for increasing exploitation of the planet's resources is now leading us to deadly and chaotic outcomes.

Actually, many wild populations of animals have either built-in biological means of regulating their numbers somewhat in response to dangerously high demands on their resources (food, breeding spaces, etc.), or can rapidly recover from a population crash resulting from starvation or other causes. If these responses did not exist the survival of species would be short, measured in years or decades rather than the thousands or millions of years that seem common. In fact, then, we should be recognizing that our own species' behaviour towards our environment is biologically atypical and we must try to understand what lies behind human avidity for over-abundance of foods, goods and space, which are leading to a world of social injustice, chaos and possible disaster.

Among our ancestors living in Nature it seems safe to assume there would have been contemplation based on lore and myth passed down orally, and sometimes from depictions on walls of stone. They were interested, as we still are, in our relationships with each other and with non-human beings, and also our relationships to sun, moon, skies, clouds, weather, mountains, lakes, the sea, trees, the very big, the very small, heat and light, cold and dark ... and the stars. But over the ages of our existence, as we gradually got some grasp of the objects and processes of Nature, much of the more spiritual essences gradually got far away from us. (I am not referring here to the demands and commands of organized religions!)

We live now in the time of greatest power to change ecosystems that extends to their appearance, structure and function, the possibility of pushing them in ways and directions according to our self-defined "needs" of the moment. We have the power to destroy any ecosystem, which means the displacement or death of many or all of its organisms, if we choose to put the economic values of some commodities in that ecosystem beyond those of all its other intrinsic attributes. The only limitation on damage is the extent of our ruthlessness.

But how do we react to threats to the environment as conservationists, when faced with a public that is not necessarily unsympathetic but is unengaged?

Conserving ecosystems

It is fitting to speak of ecosystems, because any alteration to any part of the surface of this Earth, for any reason, be it agriculture, mining, logging, industrial activity, the urban spread around cities, more roads, the making of parks out of wilderness, can only be undertaken by modification of one or more ecosystems or their contained watersheds. No person, anywhere, can avoid being located in an ecosystem, or perhaps more than one ecosystem, during every day of their lives. Everyone is domiciled in, works in and consumes food grown in an ecosystem—or perhaps in many ecosystems, some of them widely separated from each other. This is true no matter how modified ecosystems may be compared to their condition before human settlement. Each of the world's remaining primeval forests is located in, or is a part of, an entire ecosystem; such forests may themselves be whole ecosystems. A farming landscape is part of one or more ecosystems, or constitutes an artificial ecosystem replacing one or more natural ones. Cities are located in what were once natural ecosystems. If we continue at the present pace—let alone at an accelerated one—to destroy, or more severely modify, the remaining ecosystems of Nature, within the lifetimes of today's young, the knowledge that the world's ecosystems were the evolutionary theatres and experiential homes of humans for so long, will become a distant memory or only found in written records.

Those who feel uncomfortable in confronting ecologists are apt to ask, "What is it that you want to conserve?" Ecologists and conservationists may instinctively resort to the shorthand of their craft in answering: "Particular species of plants and animals, communities of organisms, watersheds, natural landscapes, ecosystems."

Many people will respond positively to proposals to save certain "threatened species," but misunderstand "communities" and "ecosystems," taking these terms to stand merely for big parklands, often with forests, streams and wildlife, where there can be hunting and fishing, mountain landscapes, spectacular coastlines and beaches, open spaces where people can camp, swim, trail-ride, hike, ski and climb. Places into which they can escape from city living and polluted air and get some sunshine and recreation in the open. Even managing to hold the present conditions steady in such places may be an admirable conservation achievement. But many of them have been formed from, or sited within, considerably altered or damaged ecosystems. And because of what they are being used for, they may even be deliberately prevented from exercising any tendency to revert to their original states. So these are not always places that rank highest in the priority lists of serious conservationists. Of course, such landscapes sometimes contain aspects or vestiges of original ecosystems, such as plant or animal species that may be endangered, rare or unique.

People may believe they empathize with conservationists, but may become uneasy and impatient when conservationists insist (gently but firmly) on defining the true meanings of species, natural communities and ecosystems, and on clarifying their use of the term landscape. "Community" and "ecosystem" are not recondite terms but are often misunderstood and misused. We should insist on their real meaning, so that people will grasp the true scope and magnitude of the conservationist perspective, which may not be modest, but may also have the ability to challenge and stimulate their imagination. Of course, there may follow more demanding questions, such as: "Surely you can't dispute that the present world—of whatever civilizations it is composed—is built on the necessity for agriculture and the establishment of villages, towns and cities, which have always required the inevitable modification, even the destruction, of 'natural' ecosystems? We have always had to alter the land surface if we are to grow crops, raise livestock, mine, selectively drain or conserve water, build towns and cities; in a word, convert all or parts of original ecosystems into 'human' environments where we can be comparatively safe from natural threats and disasters ... and where we can multiply (as religions and economics have urged us is desirable), secure in our shelter and food supply. To achieve these things we have simply had to modify ecosystems (as ecologists and conservationists call them!) in a big way to make societies and cultures possible. You surely can't argue that, in modifying ecosystems to meet societal needs, we have somehow failed to do the right thing!"

This can be difficult territory for conservationists. They can ponder why, in the course of history, a majority of people did not immediately appreciate the damage being wrought by so many of their activities and resist this damage. But it may be best just to urge that it's time for a reawakening, because it is no longer unimaginable to anticipate a tipping point beyond which Nature, as we can still know it, may altogether collapse. Evan Eisenberg (*The Ecology of Eden*) says: "If we made better use of the space and resources already at our command, we would leave more room for wilderness, which would give it a chance to function on its own." But the impact of economic forces driven by nationalistic ambitions, and the fact of our ever-increasing numbers, already mean that we can never again expect to inhabit huge samples of the primal "Garden." What we can do is find ways to retain, through conservation, more than a few poor vestiges of that pristine Nature. As the combined exaltation and apprehension of technology threatens to overwhelm us as creatures of feeling and spirit, conservation is our sole means of ensuring that we need not lose all contact with our roots.

Increasing our understanding of ecosystems through science will not diminish their status as objects of wonder and veneration, for all our studies will never quite take us to their central core. There will always be a further level of understanding. As we seem to approach the still point of the turning world, all signs are that the deeper meanings of Nature will continue to elude us and further challenge our understanding. Charles Birch has covered this comment on the nature of scientific understanding in his book, *Science and Soul.*

In promoting conservation on a world-wide basis we will be asking people everywhere to do things that society in general has never been comfortable with: curb, contain, in some cases drastically diminish the material comforts and more frivolously wasteful trappings—the toys—of its day-to-day existence. And for what? A world in which there will still remain examples of Nature to which we may devote something of our powers to meditate, to dream, to focus ourselves in asking what are we doing here and why are we doing it?

Many people nowadays appear sanguine that we will one day reach the stars. Perhaps. But it will be very terrible for those who do eventually sever contact with the essential features of ecosystems around the world. They are the places in which our brains evolved that gave us, along with language, the ability to engage in science, engineering and the arts. And the ability to ponder our existence. We need to know that examples of many of those ecosystems are still within reach, even if, in any particular ecosystem in which we happen to live or depend upon, our patch may seem to have been long separated from its beginnings and from the life and the dreams of times when we and Nature were more unified. We will need to reawaken this association between Nature and ourselves for otherwise a heedless positivism will continue to drive our technology to fatal inroads into the wondrous and still mysterious biological systems we all are heirs to.

Ownership

One of the very greatest problems now facing conservationists, who actually hope to succeed in their mission, is that the entire world is owned! Its parts are owned by individuals or families, companies or industrial enterprises—or by states or countries. How then, can we perform comprehensive programs of ecosystem conservation if owners (private or state) declare that land is theirs to do with what they want? How can migratory birds and mammals be safeguarded? And, given disputes over fishing rights in the open oceans, how is it possible to safeguard what remains of fish and whale populations?

Ownership! We have to start by asking how anyone can really be said to own land or the commodities produced on land, except perhaps the little bit of ground where they have a house, an apartment, or a mud hut. To lay claim to land can only be because it's been seized, or "given," or passed down by parents, relatives, friends or the state, or because of purchase. But from what principle does permanent ownership actually derive? This is a question many of us who believe we do "own" land will not want to visit, for it can never be truly fair that land—a limited and exhaustible commodity—has ended up in the hands of a minority. We can only hope, in the kind of world we live in, that if land continues to be considered inviolably owned to be disposed of more or less at the owner's discretion, conservationists will strive their utmost to encourage acceptance of a doctrine of generous and wise stewardship. All we can do at present, it seems, without huge amounts of money and intrusive legal, or sometimes violent, confrontations, is to try in every possible way to persuade owners to become increasingly understanding of their relationship with the earth, its uniqueness and subtlety, and that their occupancy implies responsibility, along with its privileges. They should be urged and helped to understand it is reasonable to expect them to be good and enlightened caretakers and enable others to benefit from their good fortune. If they can't be persuaded to do that, if they never come to see what the true importance of their bond with the Earth and with living Nature ought to be, the future for conservation and for the mental well-being of our species may be very bleak. Meanwhile, we should be grateful to those landowners who have already accepted their role as wise stewards and are doing everything they can to safeguard and conserve the land in their possession. Some landowners are, in fact, often protecting land and forest sites and seashores more effectively and empathically than government agencies are able to at present for many of the lands that are publicly owned.

Managing ecosystems

Don't be misled into thinking that this question has already been adequately dealt with. The conservationists' mission isn't just about learning, science, being generally well informed about the world, or articulating the issues. The conservationists' task lies in making sure that such knowledge is really understood along with eventual perils to our minds—or souls—if it is not acted upon. And in acting upon it we must have recourse to principles and laws that are crafted with greater intelligence and sensitivity regarding their application to problems of the conservation of Nature than we have at present. It's about love and respect for the Earth and its life forms. It's a question of ethics.

Nothing I have written is meant to imply that we should ignore the preservation and management of already damaged or extensively modified ecosystems, which is the state of the majority. After all, change can run all the way from slight to catastrophic to ruinous. National parks, or watersheds that feature agriculture, forestry, mining, residences, recreational areas, even townships, can still be places in which aspects of former ecosystems are readily identifiable. And badly degraded ecosystems can sometimes be brought to a state of partial recovery of their lost attributes. But even though many environmentalists labour hard and honourably to safeguard such areas from further damage, those areas can never be authentic substitutes for the originals.

Countries and states vary widely in accepting responsibility for ecosystem conservation and many environmentalists believe the future must increasingly include the ecotourism that is now so prominent as a principal component of stewardship management packages. But many ecosystems are fragile and, even if conserved, may still accumulate significant damage from visitors over time. A number of ecologists have recently been thinking that carefully selected samples of species-rich ecosystems could be best conserved by removing them from almost all interference, present or future. E. O. Wilson of Harvard University has been prominent in such proposals for setting aside areas in the tropics so as to save many species from extinction and keep such places intact for future study (which may incidentally yield new medications and foods) and, presumably, for contemplation and meditation. The basic argument is that this is a vital necessity for further understanding of the biological evolution of many species, including our own. The plan would depend on purchasing such environments, and the price would not be small! And here, again, the problem of property rights arises. If such schemes were to prove practicable they would still be only partly effective in providing conservation for many species such as those birds, mammals, fish, reptiles and even insects, all of which may spend parts of their lives in different ecosystems, sometimes separated by as much as thousands of kilometres. It may prove true (particularly for certain large and mobile mammals which historically moved over large distances within, or even between, extensive ecosystems but are now barred from normal range movements by destruction of much of their original habitat), that relief can sometimes be provided by establishing new movement corridors that allow the bypassing of destroyed forests, roads, railways, pipelines, towns and cities. But such measures will not be universal remedies for engineering ventures that cover wide distances and areas disruptive to the lives of many wildlife species.

The warming world

It should be realized, as global warming takes hold that species of plants and animals may disappear from even well-chosen conservation sites, driven out by temperatures that have exceeded their tolerance. Anticipation of such possibilities should be factored into any rational plans for large scale conservation. Much knowledge of what to expect has already accrued from experimental evidence and field observations for organisms such as trees and freshwater fish (see Chapter 4), and of course, for crops and agriculture in general.

It might be anticipated that the mobility of most animals would help them buffer the effects of rising atmospheric temperature on their distribution and range more readily than is the case for plants. But remember rainbow and brown trout ranges are confined already to those waters which do not exceed their laboratory-determined upper lethal temperatures (Chapter 4). It can therefore be confidently expected that, as global warming intensifies, the ranges of many fish species will be reshaped as they are driven towards higher latitudes and altitudes; if they survive, their range limits will certainly become more restricted. It can be similarly anticipated that marine fish, which also show great interspecific differences in the thermal regions of the seas they inhabit, will eventually display changes in their distributions. Taken as a whole, the inhabitants of warmer parts of marine and freshwater environments will experience reduction of the limits of their range while the species already confined to cooler waters will have their ranges narrowed. Such changes will have profound implications for fisheries as well as for the ecology and the very survival of many other species of plants and animals, of both terrestrial and aquatic habitats.

How to get on with ecosystem stewardship

There is little that is simply, easily or inexpensively attainable in the struggle by conservationists who came into being as an active group that was soon becoming large enough and vocal enough to influence public attitudes if things had gone a little differently. Many ecologists and conservationists can remember the period from about 1965 to 1972 when it might have been possible to circumvent much of the world-wide environmental havoc wrought in recent decades. An absence of public vision destroyed that opportunity. Now the struggle has become too often like war, which is extremely unfortunate. Ecologists, conservationists and environmentalists simply have to continue their labours until there is a public that is awakened again, as it was beginning to be thirty to forty years ago, to know and love Nature. Now the public, whatever their other aims and ambitions in life may be, simply must not allow what remains of the natural world to be lost forever.

When wide concern for any major conservation project reaches a critical threshold, so that there is a truly public determination to carry it out, then the central question will no longer be charismatic leadership, but simply how to do what governments and businesses are already challenging volunteer conservationists to do: pay for and administer the projects and perform the day-to-day tasks of conservationists in the field. Fifty years ago, most of these problems would have been the responsibilities of the qualified personnel of fully-funded departments or agencies, under federal, state or provincial directives. But today the situation is more complicated. Governments everywhere have savagely cut expenditures on public works while purporting to encourage and even fund volunteer NGOs to accept roles in what would once have been the exclusive mandate of their own (i.e. government) departments. The trouble is that these same departments, which were once much wealthier, more powerful and fully staffed, are now unable to function as they then would have. Yet they often provide funds to volunteer groups to do environmental work, while they still attempt to impose heavy-handed control over the activities of the groups' activities, even though the level of funding they disburse can be very modest. To give an example, a grant awarded to a volunteer group in Canada to perform what is perceived as a significant conservation study, or an environmental alleviation, may be of the order of one year's salary for one competent, mid-level, government department operative. Out of this, the volunteer group leaders, who usually receive no salary, and may be obliged to pay for various unfunded items out of their own pockets, are expected to plan their projects at a professionally approved standard, locate, inspire and drive themselves and other volunteers to perform the projects at a level that the department professionals themselves no longer can. Yet the latter retain the power of approval or disapproval. It is hardly necessary to declare that this situation, while accepted by dedicated volunteers as something to endure if the cause of conservation is to survive, cannot be regarded as a satisfactory permanent arrangement!

Recent studies by the University of New Brunswick's Shawn Dalton on watershed management in relation to management of the Gwynns Falls watershed in Baltimore found the strongest and most frequent interactions were between the largest, longest-established and best-endowed agencies, with city, federal and county agencies clearly dominating. These are the very agencies that would likely have dominated these activities many years ago to a much larger extent than now, though, as we see, they do in fact still impose themselves. Dalton also found the interorganizational efforts of businesses, expected to be highly significant in a large city, were actually minimal. This is likely a warning not to expect much real interest from businesses in managing public utilities!

So if we are anticipating a "modern" multi-organizational approach as a means of coming to terms with a conservation or resource-management problem we will likely find ourselves still relying on the interactive efforts of long-established state agencies. They have the experience, expertise, funds, infrastructure and authority. And they still retain some of the skilled personnel which recently-formed, and largely poorly-funded, volunteer groups will lack. Even if they are not as wealthy or powerful as they once were, they are used to dealing with each other and with the public through channels that may not be readily accessed by that public and certainly cannot be matched by volunteer groups. Perhaps the best thing volunteer groups can do is to continue insisting, as lobbyists everywhere do, that governments must not continue to withdraw from conservation matters. For in fact, if governments eventually have to abandon this kind of activity to volunteers, no matter how enthusiastic and well-intentioned the volunteers may be, and to the business sector, conservation will perish as a serious form of societal activity.

What is clear about how these various scenarios concerning conservation projects are being performed is that the situation is stopgap, basically unsatisfactory and needs rethinking from top to bottom.

So conservationists must find appropriate heroes who can persuade an entire world of the validity of their mission. And this is not a task to be postponed for some golden future when our world will have largely renounced its curses of erosive competitiveness, force and bloodshed, when there will supposedly be fairer distribution of wealth and universally better attitudes. However jolting the various prognoses of the future may be to those who are trying to remain simultaneously aware yet calm, learning to do the right things for Earth's ecosystems and areas of wilderness is now simply a part—and a very large part—of getting to a better world, in which there will be universal respect for plants and animals—our fellow-beings—and for the integrity of their environments.

Only connect!

The beauty and splendours of Nature can still catch us in strange and unexpected ways. Consider the experience of Lawrence Osborne, who recently wrote in *The New Yorker* of his travels in the dense jungles of New Guinea (*Strangers in the Forest*). He and his companions were newcomers to such alien and demanding environments, which required considerable adaptive effort on their part. Yet, "There was, in our grisly slog through the forest, a shimmering animal joy—not pleasure or excitement—that deepened each day." He recalls "swimming in a sinisterly beautiful river" with a white beach, and in which "the water was deep and opaquely cold and there was a threat of crocodiles." Another in Osborne's party called the forest "the most beautiful place I've ever seen." Later, Osborne compared notes with another who had accompanied him and asked for his impressions. "Like a dream," he said, "a complete dream." And he recalled, "That beautiful river ..." And he told how he and his own companion were "having quite a hard time adjusting. We suddenly realize how weird and noisy our culture is. We miss the closeness with nature." Osborne also cites a 1961 visitor to the same forests who wrote of the strange, seeming-mediocrity of all civilization, even art, music and philosophy, when he came out of the New Guinea bush. That visitor speculated that it might have been the remoteness of the things and influences "of civilization from the real nature of man and his natural world that make them appear flat and unreal." It seems that sophisticated, modern town-dwelling humans can still be reached and re-sensitized to Nature in ways, and at levels, they could not readily have anticipated.

It would seem not unacceptable imagery to think of the Earth of the past as a spatial object resembling a vast green and blue, white-streaked opal, a shimmering object of unique beauty and eternal value. In present days, astronauts have had views of what remains of that object—the first humans in history to do so! It was, and still is, there for us to revere and contemplate, study and try to understand. But those of us who now call ourselves conservationists are already feeling something like curators of the remains of a priceless gem, concerned to guard against its further damage that will render it unrecognizable.

CHAPTER 8

BUILDING ENVIRONMENTAL
PERSPECTIVE

A permanent enhancement of one's perspective is probably the most important gain from having participated in conservation projects with the accompanying use of ecological analysis. The kind of work done in two such projects and the resulting understanding is described in Chapters 5 and 6.

I have given some account of the place where I live (a river-lake environment, described in Chapter 6), a brief sketch of its human demographics and of the reasons why people choose to live in this particular watershed. The reasons are very different from what they were nearly two centuries ago when the area was just beginning to become an agricultural and forestry location. Nowadays people come here for various reasons. Many who live here permanently may commute long daily distances. Others spend holidays or retire here. But one thing is clear as discussed earlier, the system is not completely safeguarded against repetitions of the environmental damage of twenty years earlier, even though its causes and the means by which it could again become seriously threatened are now understood and potentially preventable. Armed with the understanding we now have I would always find it easier to evaluate the impact of a period of human use and occupancy on a settlement of any river-lake system, whatever its size, and regardless of whether the impacts came from industrial practices or from causes similar to those of twenty years earlier In a word, we have not just acquired knowledge of how to manage a specific impacted system, we now understand much better how to investigate a great range of watersheds.

Similarly, the knowledge of what happened to a river as a result of its being polluted by zinc (Chapter 5) means, depending on the severity of the pollution, that one can anticipate in broad terms what will become of a great number of metal-polluted systems. Of course, the specific biological effects on a system's organisms will depend on the particular metal involved. Lead, for instance, mercury and uranium in a human environment have long been understood as injurious to human health. But almost any ecological survey of such pollutants will likely resemble what is described in Chapters 5 and 6, though the technical tools may differ from those we used. Take, for example, the highly dangerous effects of uranium mining should its wastes enter a watershed, as has been a great concern of many people living in the Canaan-Washademoak following events that suggested such an outcome in this part of the Saint John River system. Most people know of the harrowing experiences that followed the notorious failures of nuclear plants in places like Three Mile Island and Chernobyl, but many fewer will have thought much about the comparatively small amounts, or much more gradual effects, of the entry of uranium mine wastes into a river or a lake.

It is easy to claim that modern technology is designed to ensure radioactive wastes can be safely contained, but wary environmentalists will picture the disasters that could accompany the release or leakage of even traces into already inhabited places. People living in such proximity of the effects of radioactivity are justified in being appalled at such possibilities. Excuses or explanations always arise when there are profits to be made. But any mining in the precincts of large rivers will put widely populated sections of the human environment at tremendous risk of permanent health damage.

The importance and relevance of these apprehensions can be well understood when one reflects on the example of the zinc pollution of the Molonglo River (Chapter 5). How easily unconsolidated wastes were swept away from the Lake George Mine to be distributed along much of the river's course for about 50 km, there to lodge for more than 60 years following, and everywhere to contribute permanently to the waterway's toxification. Imagine how dire and enduring the effects to human health would be if the zinc had been uranium. Our evidence from two river-lake systems tells us what we could expect. To be able to argue thus from the already existing evidence is much more economical and certain than being forced to repeat grievous, perhaps deadly, errors with extremely long-term consequences.

Actual experience with field studies of damage to ecosystems really can provide knowledge that builds towards a more reliable base for subsequent planning and actions without always needing to start entirely from scratch in every single case of environmental threat or damage where conservation may be called for. But forging ecological tools and learning how to view new conservation problems in the light of past ones is only part of the struggle to attain a universal conservation perspective. The most important problem, and the hardest to solve, is how to get people to embrace that universal perspective.

We can perhaps acquire this only in one way: by discovering people who are able to bring transformative and transcendent rhetoric to bear on the cause and necessity for conservation in a manner that will ignite wide public interest, opinion and action. Many people with great concern for Nature warn, in stating the case for its conservation, that one must at all costs avoid becoming too negative. If audiences become alarmed by too many concrete examples that try to illustrate how enormously dangerous to the future of the planet are many things that humans have already done and continue to do, then the attention of most people will dwindle. They will be turned off by the notion that the cause is hopeless, there is nothing to be done and they might just as well shrug and enjoy their lives until the end. From their initial interest and idealism they may simply slump down into indifference and passivity. Perhaps. But this advice may be too easily accepted.

If one examines many famous historical speeches that successfully urged large audiences to confront and actively struggle against even seemingly deadly challenges to their existence it is clear that people may rise to the occasion. If people can become convinced that the depth of a disaster may be gravely damaging to the quality of their lives they may accept the challenge of countering or preventing it and experience the benefits and joy of prevailing against it. These points are conspicuously present in the WWII speeches of Winston Churchill. He did not flinch from explaining bluntly and in considerable detail to the widest possible audiences that terrible damage had been done to the Allies' cause and that survival demanded extravagant sacrifices. But there was always a special and especially attractive tension in the structure of his speeches. They managed to overturn the actual gloom and doom that threatened by then stressing the righteousness and idealism of the Allied cause, and they spoke unfailingly of the enormous rewards that would result: not rewards of money, position or status, but rather the maintenance of that quality of life I have referred to. This was how he managed to set the terms, indeed the limits, of a challenge and accepted that challenge in an inspirational way on behalf of audiences of millions. Read his speeches today. The words still live.

But Churchill wrote of resistance and victories in war. Is his type of rhetoric suitable for conservationists? I think some versions of it may be, if only because we are in for a protracted and global struggle requiring sacrifice, and what could be more idealistic and engaging than regarding this struggle as something we must win, as in a war? A war in which victory will lead to the saving of Nature.

Churchill insisted that to lose the war was unthinkable. Surely conservationists should adopt this attitude in their struggle. Somehow, from somewhere, conservationists have got to find and employ speakers who must be prophets and idealists. Speakers who will arouse not just the audiences of the convinced, but also shore up the nerve and the faltering enthusiasm and idealism of those who are sympathetic and empathic, but whose vision does not yet quite encompass the horrors of a permanently neglected world, nor the belief that it must and can be saved, if enough people insist that it be saved.

CHAPTER 9

SUSTAINABILITY IN SOCIETIES OF THE FUTURE

In any attempt to consider what can and should be done about world conservation we come at last to what must be, for our species, the ultimate conservation questions. I mean the problem of how to view the past, present and possible future state and welfare of human societies. How did they get the way they are, how do they function individually and en masse, and what will their future forms take? What I term a society can be as small as a village, as extensive as a populated countryside or a city, or as large as a nation. While all societies change in many ways over time, they often retain recognizable features and qualities that endure for long periods. In the contemporary world we can hope that the societies we inhabit must increasingly endeavour to ensure that their citizens have access to the necessary material conditions of life which, at their minimum, reduce to food, water, clothing, shelter, health care, education and civil rights. But for these conditions to be met and maintained we must plan and work for them in rapidly changing, violent and chaotic times. To succeed we must recognize various inescapable facts of existence that will deeply influence the types and patterns of societies to come. Primary shaping forces will be derived, more generally than ever before, from increasing regional pressures on populations. Agricultural land, forests, fisheries, mineral reserves, water and power supplies will change and shrink. And both the short-term and long-term effects of pollution and atmospheric warming at both local and global levels will increasingly assert themselves.

It is possible, of course, that greed, competitiveness, tribalism and brutality are hard-wired into humans, making it impossible to ever achieve a truly viable global nexus of societies. That is what John

Gray in his book, *Straw Dogs*, suggests, forecasting a doom-laden human future.

However, just in case some people cannot really conceive of what it could be like in a world where land, food, water and space were really limiting they should note the examples of the isolated Easter Island and Tikopia Island in the South Pacific. They serve as terrible but genuine examples of the tenacity of the human spirit in the face of extreme environmental poverty, but also of the catastrophic outcomes. In Easter Island (area 66 square miles) the population relentlessly declined in the course of a millennium as its natural resources were exhausted. Tikopia, a tiny island of only 2 square miles, with very few natural resources, managed to maintain a population during a comparable time period by scrupulous use of all available space for plant growth, capture of marine fish and avoidance of internal conflict. However the struggle of its people to survive was constant and involved infanticide as a population control.

The fate of these islands can serve as small-scale "demonstrations" of societal survival under fixed conditions of extremely limited supplies of water, food and space. Because these conditions were permanently limited they differ from the awful, forcibly crowded conditions imposed by wars and genocide. But they bring us face to face with what life would be like as long-term survival under desperate conditions.

Several kinds of societies may be thought of as sustainable; some of them may change a lot over time, ranging up and down the historical scale between small and simply organized to large and complex. Indeed, they may differ in many ways, and even individual societies (or nations) that are large and complex may include within themselves great regional differences in living conditions (e.g. India, China, Brazil, the U.S. and Russia). But any particular society, whether financially rich or poor, may be not just sustainable, but essentially self-sustainable. This means that such a society may, or could, survive in some manner without the need to trade or interact with others. For example, even monetarily poor societies such as those of Amazonian natives, or the Inuit, have been able to find and subsist on just the bare necessities of food, water and shelter. Aboriginal Australians survived at low population densities, until less than two centuries ago, under a variety of climatic conditions, sometimes very harsh, for tens of thousands of years. There are, of course, some societies, whose basic access to food and water was once barely enough for their survival, but which have managed to become comparatively rich by finding and selling a globally prized natural resource, such as oil, to more complex, wealthier societies.

Another group of societies, intrinsically self-sustainable, includes those that can not only meet their own needs for food, water and shelter, but may be able to export a little or even a lot of raw materials, food products or sometimes manufactured goods of more or less complex design, which may or may not require high technology in their production. Such societies as these, if they possess enough agricultural land, raw materials, power sources and industrial potential, may continue to nourish and shelter themselves while they morph into societies with the capacity to satisfy not only their own physical ambitions, but to earn appreciable financial wealth from a variety of exports. If they are big enough they can, of course, also become expansionist, powerful in global politics, even hegemonic military powers dangerous to the world.

The great oxymoron

In most of the 20th century it was popularly held that, regardless of huge increases in world population, the ability to feed, clothe and house people would "always" be achievable, thus relegating the spectre of a Malthusian future to the realm of failed hypotheses. Ever-increasing industrial production of goods and more "efficient" agricultural and economic practices were the supposed magic transformative processes of societies. To be sure, there have been disproportionately vast increases in the monetary wealth of certain already industrialized states that have grown increasingly more dominant internationally. And it is evidently their example that has given rise to the concept of "sustainable development." But this concept is ridiculous, because it asserts that a general condition of constant enhancement of financial wealth, health, comfort, material plenty, plus many other tangible and intangible goods and services, along with growing production in agriculture, forestry, mining, power generation and industry can all occur and be maintained, even in conditions of continuous population increase. Such increases always involve use of more energy, raw materials and fertile soil, commodities that are constantly decreasing in obtainable or reusable amounts. It therefore defies reason to believe that material and/or economic development can be sustainable in any long-term and true sense. But whatever its basic absurdity, sustainable development has become a part of the "technobabble" and the mantra that businesses use in hoping that their more extravagantly far-fetched schemes will curry favour with society at large—or at least not attract public hostility. It is truly unfortunate that the term has fallen into such widespread and uncritical use. Consider the following: Jared Diamond in *Collapse* described past and present conditions in Australia and explained that, despite the arid conditions, including low and unreliable rainfall and soils that are on average of very low fertility, most Australians, for well over a century, thought of the country, because of its giant area, as a place of great potential in primary production, in which manufacture of many goods could readily outstrip population growth, that exports would come to constantly surpass imports and that the wealth of its citizens would steadily improve. Until the very recent past, many Australians would have considered the country able to be self-supporting in food for a population of fifty million, while still having wool, wheat and minerals as major exports. Now Australians are starting to understand the country's fragility, with an agriculture based on scarce water and soils that require heavy inputs of chemical fertilizers to remain productive. Farming is expensive and will get more so. The vast mineral resources may partly sustain the economy for a long time, but with declining wool and wheat exports the future is uncertain. Diamond has speculated that if the population much exceeds its present level it may end up as a net importer of food. He notes that many influential political and business leaders still dream of the fifty million population, whereas less than half the present twenty million would be more realistic if the country were to continue as anything like self-sustaining. In a recent *Globe and Mail* article Peter Russell of the University of Toronto observed: "Australia has gone from being a British farm to an industrial trading company in Asia." This pinpoints what has happened to the country and that its future does not lie in the earlier era of exports of primary produce. It can still export its rich mineral deposits, but this is emptying a bank rather than creating a constant source of income.

Turning to China, Gwynne Dyer has recently stressed the fallacy of many of its blossoming ambitions, and of those of other large countries, to be as affluent per capita as the present (or recent) U.S. Dyer notes that "the predicted development of China by 2050 (and the comparable growth of India, Brazil and Russia) would raise the share of the human race living in high-consumption industrial economies to (more than) eight billion by that time," and that this will amount to a total human pressure on the environment that is twenty-five times greater than it was at the end of WWII. Other similar examples of countries hoping to rival the U.S. in the per capita use of energy, materials and food, are given by Christopher Flavin and Gary Gardner in *The State of the World, 2006*. These authors also note, however, that both China and India are now showing interest in "leapfrogging" Western countries in developing more economical uses of energy, water and transportation technologies, which may signify a more realistic and progressive attitude towards some aspects of the future world than is common. However, such economies would only partly offset the eventual impacts of these giant states on the rest of the world and, of course, on their own citizens.

What it takes to be a sustainable society

To view some of the points already raised from a different, more fundamental perspective, some other objective measure than their dollar earnings is needed against which to understand significant differences between societies' most important resources and also in understanding the true meaning of wealth and its occurrence. In 1776, Adam Smith, founder of modern economics, pointed out in *The Wealth of Nations* that the rapid production of financial wealth in America was basically the result of early European settlers coming to a country of vast natural resources, bringing with them from Europe technical knowledge resulting from the early stages of the Industrial Revolution and the burgeoning intellectual skills of the Enlightenment. These were ready-made advantages for the rapid production of material plenty and financial wealth. Today we can use this vision of assured wealth in understanding that nothing is more important for societal self-sustainability than a country's biological productivity. Thus, more than 30 years ago, S. R. Eyre offered in *The Real Wealth of Nations* a masterly analysis that, in broad pattern, would, I believe, still hold today. Eyre estimated and listed natural endowments of 147 nations in terms of their potential tonnes per capita net primary productivity (PPCNPP), the basic necessity of self-sustainability. Among the 27 least endowed (in this regard) the range was from less than 1 tonne to more than 9 tonnes PPCNPP. For the 23 best endowed nations the range was from 200 to more than 5,500 tonnes PPCNPP!

Among the worst endowed nations, those predominating were heavily populated European countries, dry countries of the Middle East and southwest Asia, heavily populated and formerly forested countries in southern and eastern Asia, and heavily populated islands in the West Indies. The UK, though highly industrialized, technologically advanced and militarily powerful was, and would still be, near the mid-range of the worst endowed societies. Yet the UK is popularly considered, and indeed its people think of themselves, as one of the world's more affluent, sustainable societies. Its dense population is well-fed and housed, has abundant food and consumer goods, high literacy, universal education, health care, high longevity, low infant mortality, a stable population. But the UK's material quality of life is only sustainable as a result of its earnings through export of manufactured goods and its sales of various technical and financial services, made possible by complicated international trade and money management. Eyre pointed out, though it had been obvious since well before WWII, that for the UK to maintain its population's needs and well-being, it had to import at least half its lumber and wheat, a fifth of its other grains, more than half its sugar, butter, mutton, lamb, bacon and ham. It received vegetables and fruits from a variety of other countries. Domestic wood and leather supplies were also far below the nation's needs. Coffee, tea and natural fibres came from abroad, as did agricultural fertilizers. However, the UK has long managed to maintain its society in a condition of relative comfort and ease, but without huge internal reorganization, could not continue to be a "sustainable" society. It depends on an intricate pattern of international trade and services that it has established over an extended historical period. The UK's "prosperity" results from a mixture of industry, politics, manipulative economics and power. As a matter of note, the UK's economic condition had become quite precarious in 1975, following a post-WWII peak. It was the use of North Sea oil that rescued it from this condition and restored its wealth. But this resource is no longer at the same level of significance. Is the UK, therefore, actually a "fragile" society? The answer depends on how long and effectively it can continue, along with similar countries, to broker profitable deals with other societies, many of which, though not usually considered wealthy, may be both self-sustainable and also exporters of goods or services.

Thirty years ago India was another nation whose PPCNPP was near the mid-range of the worst endowed list given by Eyre. The population density resembled that of the UK, but its per capita income was only about one-thirtieth. Today its population has grown by about a third, and its per capita income has increased to about one tenth that of the UK because of the nation's large increase in industrial capacity and business, especially in high tech electronic products and services. India had been compelled to produce cash crops to buy fertilizer and machinery for its basic national food supply. It may be easier for India these days, with more money at its disposal, to pay for its basic crops, but with its great population increase it has probably made unsatisfactory overall progress in agriculture. About one-third of the population still lives in great poverty. However, during its period of economic growth, India has in some ways remained closer to being truly self-sustaining in agricultural production than the UK.

To note again: the inhabitants of the UK have enjoyed far higher general standards of living than those in India because of the way the UK continues to conduct its economic affairs. But if it were to lose much of its powerful business and industrial export strength, and had reduced access to world supplies of food, its living standard would come to resemble that of contemporary India. That is, it would become a society with a comparatively small, wealthy elite, a modestly comfortable middle class and a large, spectacularly impoverished, underclass, much as it had become in the earlier stages of the Industrial Revolution. Comparable arguments can be applied to various other Western European nations. Certainly, the present economic standards of most of these societies would be otherwise, if not for complex economic arrangements that have little to do with natural resources or agricultural capability.

About self-sustainable societies

It is the relentless application of certain forms of technology in modern times in the U.S. and other industrial countries, coupled with their acquired dominance in global economic affairs that has led to mass production of goods that outstrips all basic needs of comfortable living, leading to the modern cult of consumerism. This cult has led to proportionately outrageous profits for some and extremely wasteful, resource-squandering styles of living for the many. A belief underlying these ways of living – astonishingly widely held – is that any society which really wants to can attain high industrial production levels, maintainable for very long periods, and live with the reckless wastage that particularly characterizes nations of Western Europe, the U.S., Canada and Australia. This belief is now driving the contemporary industrial economies of the giant populations of India and China. However, as already noted, their most pervasive ambitions will not be realizable because the global limits to what they would need will not let them. Many still adhere to the notion that human ingenuity and entrepreneurial efforts – which have thus far affected our social evolution to the extent that we are not merely reactive to our conditions of existence but can be active, even pro-active – will still manage to rescue us from forced modifications, even breakdowns, of our societies. But there seems no real evidence either that societies' end games will be played out in terms of technological magic, or that Malthus' predictions have been anything more than delayed in their outcome. Thus Gwynne Dyer ("How long can the world feed itself?") recently noted that "for the sixth time in the last seven years, the human race will grow less food that it eats this year" (2006). He points out that it is only by consuming accumulated food stocks since 1999 that world shortages have been avoided. That period is now over.

This may seem a hard view of societal possibilities, because financially deprived countries feel great satisfaction in attaining greater levels of material prosperity, and may experience despair when physical circumstances of existence thwart their plans. On the other hand, if they could moderate their drive to obtain what are commonly perceived as the goods of society they could probably still attain a level of physical ease, health, social justice and well-being that would continue to serve them well when presently financially richer nations may be undergoing losses, pain and confusion.

Altogether, though many wealthy societies can be viewed as sustainable as long as they continue economic arrangements that enable them to maintain their living standards, these standards seem likely to be better assured among those that have the actual potential for self-sustainability, like contemporary Canada. Of course, any concept here of comfortable living standards should not be confused with abundance of consumer goods, food produced and consumed far in excess of genuine requirements, or of serious waste of energy (from whatever source).

In fact, it is a bitter irony that few of the simple, self-sustainable, contemporary societies seem able to resist the blandishments or pressures of consumerism. Suppose, however, that some such societies really steel their will, and ask themselves what it would take to ensure self-sustainability while enhancing whatever is needed to arrive at a secure, healthy, well-nourished condition without selling off, at cruelly cheap prices, whatever basic attributes they possess. Then they could make their sustainable life one that was truly their own and genuinely estimable. For the possibility of self-sustainability is tantalizing, since a society or nation that has the basic facilities to achieve it – to "go it alone," even if for political reasons it may find this difficult – has at least the prospect of productivity adequate for its needs, and can try to stand aloof from intersocietal trade deals of dubious benefit. But how can those financially challenged societies, which nevertheless have abundant resources or commodities, that are attractive to more powerful societies, insist on prices that are "fair"? Perhaps by simply insisting on them, and calling upon various powerful members of the international community to back them up in the interests of global justice – or, if you like, balance of powers. It can be done, or many oil-rich but militarily weak countries would have been completely looted long since for their oil by more powerful societies.

Today, even a society like Canada should not assume that its present potential for self-sustainability will last forever. Its minerals, which are currently exported in large amounts to more industrialized countries as accepted necessities of their national industries, will eventually be depleted. As national conservation measures for Canada it would also be prudent if more ways could be found to recycle the minerals that it mines and uses in its domestic economy. If Canada fails to implement much better management of its forests and fisheries, all of these formerly wondrous sources of natural wealth will fail disastrously. Canada could also run out of fossil fuels, which might eventually force it into adoption of nuclear energy on a large scale because it has failed to explore alternative ways of harnessing power. The nuclear option is, at the very best, a choice with many hazards.

The truly self-sustainable society of the future, which also anticipates a life of benefits and advantages for its citizens, will need to attend to many things in order to maintain this quality of life for them. These things can be listed, though it may, of course, prove dauntingly difficult to maintain their influence – especially in societies whose existence and structure may, in themselves, be largely the results not of planning but of inescapable global environmental circumstances. Every effort must be made to maintain population density at a level that does not exceed a society's ability to provide adequate food, shelter, health care, etc. There must be environmental protection against pollution and other forms of degrading practices. Land cultivated for farming or for use in forestry will have to be carefully managed and conserved. Only rarely and in very special instances should such land be added to, and unused or damaged land should be reclaimed and allowed, or aided, to return to its original state as far as possible. Urban spread over prime agricultural land or wetland should be curbed by law. Water resources must be conserved and kept pure. Organic wastes will be recycled or used as energy sources. Minerals will be conserved and recycled as much as possible and prevented from becoming toxic wastes. Sun, wind, geothermal power and tides will be tried in various combinations to produce or augment domestic and industrial power.

In addition to this demanding list, the self-sustainable society will produce as much of its food within its own boundaries as possible. Its trade or exchange of materials, goods, energy, services and the employment of workers will all be performed with maximum concern for the well-being of those employed, and will include free and equal public access to the best available health care.

On top of everything will be education, as the major cornerstone of all successful endeavours to reach and maintain societal self-sustainability. This need not imply mimicry of every facet of high technology educational opportunity, as found in the most technically sophisticated societies, but rather to maximize literacy and educational standards so that citizens will be able to participate fully in the organization and prosecution of their civic opportunities and responsibilities.

There is one more point that is the most important of all. This world, no matter how modified it has become, began as a series of ecosystems that occurred in a great variety of forms. The forms contained innumerable communities of living creatures and although the lineaments of the ecosystems shifted in form over time there was in many of them a consistent, recognizable structure for very long periods. We have now contrived to destroy, or profoundly modify, many of these ecosystems, either by destroying them entirely or by damaging some of their great components. For example, we are hugely changing ocean ecosystems through overfishing. We have destroyed the majority of natural forests, replacing them with forests of vastly reduced value and variety. These sad losses and many others are all working against our present ability to maintain the world's productive capacity. They also present any society that reaches for independent sustainability with an added obstacle.

Is it possible for the citizenry of small, independent, essentially self-sustaining societies that lack large natural resources and heavy industrial technology to become financially affluent in the contemporary world? A recent example to provide a "yes" has been Ireland, a society that was already able to feed and house itself but had very limited capacity to develop industrial products for significant export markets. Yet, in a single generation, Ireland transformed itself by use of its principal national asset—an educated, literate populace—into a "service oriented" economy with modern electronic communication systems and skills as its principal export products. If Ireland should lose this "industry" by being outperformed by other societies it would, of course, revert to being what it was before: a small, not very rich, but largely self-sustainable society. Compare its performance and essential self-sustainability with that of Singapore, which is also a small, technically proficient state using imported raw materials of other societies to manufacture export products by employing the education and ingenuity of its technical workforce.

Both Ireland and Singapore may be justly proud of their achievements, but in their economics they are stretching the game they play to its logical limits. They are living on their wits! They may survive thus, enjoying high living standards for a long time, collapse gradually or even quite rapidly, as other centres of production of comparable goods and services come on line to compete with them. However, Singapore, with a population of 4.4 million in 633 sq km must import much food, whereas Ireland, with 4 million people in 70,000 sq km can feed its society. Contemporary Singapore and Ireland each learned how to market their products or services with great skill. But if Singapore's industrial production or export capacity should fail, the result could be crushing, since it is a city-state on a small island and probably could not nourish its population.

We must note, though, that Ireland may be losing its unique position in providing the kinds of services that are its profitable products, which are now being offered at lower prices by other, poorer, countries.

The natural history and future of sustainable societies

What, then, to make of the concept of sustainability? For its assurance of a secure future, any society that is already self-sustainable should make every effort to remain so, even if it also engages in trade with other societies. Particularly is this true of smaller self-sustainable countries or those that have but limited supplies of raw materials, food, energy, etc. to offer the global market. Because if they begin to count financial gain from exports and, as a result, experience rapid population increase accompanied by failure of agriculture to meet growing demands, they will likely have inadequate funds to buy enough food or other basic necessities for a sustainable life of good quality. Countries that are self-sustainable and have rich natural resources and perhaps some industrial export capacity, but little or no military, business or political power on the world stage, are especially vulnerable to political pressure, intimidation or threat from more militarily powerful or influential societies. It was partly to avoid this scenario in earlier days that led Egypt, Greece and Rome, and more recently England, France, Spain and Holland—all initially societies of relatively small size—to gain control, historically, of the economies of many other societies, persuading or compelling them to neglect the internal conditions that could previously allow them to be self-sustaining. This may also lie behind the recent and current attempts by the U.S. to build a world empire by military pre-eminence. History is a tale of exploitation with many examples of the ways that initially vulnerable societies have become powerful, while simultaneously rendering other vulnerable societies weaker by managing to appropriate their natural assets, often even causing them terminal harm in the process.

The question, then, ought to be how essentially self-sustaining societies, without military or political power, can somehow avoid or limit their movement away from self-sustainability, so that their survival and self-respect can be assured. I am not thinking of societies of homeless beggars in rags, self-sustainable only in terms of a minimal sustenance, an infrequent and undependable filling of their food bowls! We have got to get to the point where people, functioning at the very least level of existence we can call human, are incomparably better off than that. Any sort of sustainable society should be one in which its people have the right to food of sufficient amount and quality and the other basic attributes listed earlier. They must have the physical and legal requirements to live a good life without the absolute necessity to engage in trade arrangements that may eventually prove ruinous by stripping them of their resources.

Of course, we live in a world of varied physical and biological conditions; local famines, natural disasters and epidemic diseases are sometimes uncontrollable. Climate and weather can enormously influence needs in food, shelter, clothes and energy. A tsunami or an earthquake can lay waste instantly to a large populated area. On the other hand, to live in a salubrious climate may be at least part-compensation for less income or goods. And in general there doesn't need to be complete universal income parity; a good and long life can be lived without frills. In the long run, societies should strive, above all, to secure a self-sustainability that is safeguarded as well as possible against economic misadventures. However, even complexly sustainable heavily armed societies with large populations, need to be safeguarded against catastrophic loss of their more essential trade arrangements, which may render them fragile in their political structure, then become mendicants and eventually militarily dangerous, because while growing desperate they may still be powerfully armed. In other words, not only might their own populations suffer, they might extend the suffering to others.

The best thing to be hoped and aimed for would be as many self-sustainable societies as possible interacting as a global whole. Then, provided the lure of financial wealth did not threaten the integrity of their self-sustainability, they could be part of a nexus of sustainable nations. Such a world would be quite different from today's, which is driven by those who think overwhelmingly of free-trade economics and of cutthroat competition in the marketplace. So where should we look for the innumerable acts of imagination, creative and humane statecraft and planning, and unprecedented honesty and generosity in international dealings that will bring that world into being? What human institution is best fitted to generate negotiations? The United Nations? Only if it can somehow acquire detachment from the present economic and military powers bent on exercising their urge to preserve the regional economic disparities that enable them to maximize benefits under their conception of a free market system. A system, incidentally, that features massive competition between trading blocs: a kind of arms race in production of consumer goods! Nevertheless, though one would not have supposed it a few years ago, there do seem to be a few signs that G8 (and perhaps soon G20) nations may at last be getting ready to face some of their true international moral obligations in areas of human welfare. If there is enough idealism still left in this world a global network of sustainable societies, many of which are self–sustainable, may yet eventuate. Its occurrence and existence would, of course, presuppose a new phase of social evolution that transcends the brute force economics, military confrontations, secular, religious and ethnic intolerances and political ideologies that colour today's relations between, and within, societies. Can we get there in a world in which too many societies choose the military option to advance their positions, and in which ever more societies are incorrigibly simplistic in the cause of boosting their nationalism?

First we must come to terms with some realities. If the world awakens to the desirability of as many of its societies becoming as self-sustainable as possible, this aim will have to be applied to complexly organized societies as well as the simpler ones. As a single example, consider again the UK. It could grow many more high-yield crops to be used directly as food for its population, cutting down drastically on beef and sheep which are very wasteful users of food energy. No doubt such measures would be unpopular in a nation accustomed to a high meat diet. And, even if successful, they would impact powerfully on the pleasantly park-like, and much cherished, appearance of the countryside. The UK population would undoubtedly find that even a partial revamping of its countryside and way of life might seem an impossibly daunting challenge. Yet it, and many other similarly-based societies, may soon be forced to consider how these things may be done.

Let us suppose that we really want, and are at last determined to achieve, a world composed of societies in each of which a good life is considered a right and in which the species as a whole can have dreams worthy of its ideals rather than its lusts for money and power. Then the concept of the essentially unregulated global economy will have to go. Otherwise, we will soon see a competitive North America desperately pitted against Japan and Europe, and a little later against China and India. In fact, the beginnings of this are already visible in terms of World Trade Organization talks dealing with agriculture; all these countries, and Canada, are fighting to see who can give the greatest "hidden" subsidies to their farmers. If these processes continue they will undoubtedly result in ruinous wastage of resources and more pollution and global warming—all in the cause of unbridled consumerism. And this would become an economic orgy that would destroy many societies with little to trade and no means to buy what they need. At best, they would be beggared, tossed the bones from the tables of the rich. And the world would be an increasingly dangerous place.

Many societies appear presently faced by the dual burdens of inadequate food supplies and seemingly uncheckable population growth. This enormous latter problem, only intermittently acknowledged over the past two or three generations, as human numbers raced from 2 billion to 6.5 billion (the last billion coming in the past decade), will derail our best-planned schemes for societal rescue. For the homely but potent message proclaimed by Paul and Ann Ehrlich, for more than a generation, will remain to remind us how preposterous is our present world tolerance of this matter: two hundred thousand more mouths to feed with every passing day! If we, a "we" that includes all the religions, political and economic systems of the world, continue to ignore this we are not just stubborn and blind, we are clearly crazy. And we will abundantly deserve our neo-Malthusian fate.

Neither politics nor planning, just the pressure of a narrowing world

A world of the future, when its peoples finally grasp the need for tolerance and peace, will not, cannot, be a place in which dreams of ever-increasing material plenty can be manifested. Limits on space, energy and food production will curb any such scenario. And, in seeking to offset the more extravagant ambitions that could otherwise burden societies with impossible dreams, serious planners will be forced to hew to a far more modest and realistic vision of the future of societies. That future, if it is to have any prospect of global stability, will feature a nexus of human societies, where some will have given and others received, but in which the watchword will be that all valid and viable societies will have moved as much as possible in the direction of self-sustainability. Of course many societies will not have achieved self-sustainability; there will still be trade, commerce, movement of people and products between societies and nations. This will be a world of realities, whose lineaments are not yet properly in focus. It will be a world of limits rather than one of unbounded possibilities. But it may be far closer to a humane and happy world than we have at present a right to expect. It will be a world that will largely define itself, because the conditions that so many have allowed themselves to be persuaded can be universally obtained will be found out of reach. And the human species must one day acknowledge that it has to live with this. Or perish.

The main point in contemplating this world of future societies and how it may come into being is that it enables people to understand that materials and conditions will fashion its form in a much more inexorable manner than ever before in history. We may grow to accept its coming and begin to plan for it, so as to make its coming easier, instead of continuing to insist that it cannot, must not, be allowed to come.

In this chapter I have assumed that simpler, essentially more general and irresistible forces than political stances or machinations may determine the future of societies. The suggestion is that, given the world we already inhabit, with the conditions under which we already operate, humans have arrived at a point in their history when the courses of further civilized societal restructuring are already limited. Of course, conflicts could continue and worsen, and certainly we could waste our remaining assets and destroy our environment, and probably ourselves. The societal mix of the future will result from the remaining resources with which our ability to modify or extemporize will be constantly diminishing. The economic theories and political forces at work in the long run, or perhaps the fairly short run, may not bear much resemblance to those the world has previously accepted and acted on.

Subsequently

After I wrote what were intended as the last lines of this chapter the devastating and portentous events in the financial and investment systems of the U.S. occurred, followed by similar disasters in many other countries around the world. I cannot, of course, comment on the range of economic causes that threatens so many. But because of them, several matters of which I wrote have increasing global significance. Many international observers of the situation have already suggested that, however things play out, the U.S. may already have passed the pinnacle of its much heralded world leadership, at least in terms of wealth and in its attempts to influence the political ideologies of so many other societies, a USA that is permanently weakened in wealth and politics, while still possessing the world's most powerful military. In such circumstances everything could point to dangers for all. On the other hand it is quite fascinating to me that President Obama and many of his supporters are looking with fresh and very critical eyes at NAFTA. They clearly seem to be considering a newly self-sustaining U.S. To me this is a potent sign when it features the world's most powerful society in such a light. And it strongly reinforces the idea of future societies that I first put forward in 1991 in *Preparing for a Sustainable Society* as part of a proceedings published by Ryerson University.

As far as the details of what will happen to other societies as they become afflicted by the same condition of economic collapse, I have little insight. Many of the things I have outlined in this chapter may be influenced in ways no one can possibly foresee. For instance, has Ireland sustained a fatal blow or can the Celtic Tiger at least partially recover from its dramatic collapse? The recent tragedy of Japan caused both by natural disasters and man-made technology is an example which will have effects around the world, both economic and environmental, to say nothing of the human costs.

Thus all bets are off. Except the principles of the desirability of sustainable societies should more than ever be the basis for a world of the future in which failure of one state cannot be allowed to lead all-too-simply to further failures of others linked in ways to those that are already failing. More than ever, though, I believe that the time of the self-sustainable society will dawn.

CHAPTER 10

THE EFFECTS OF BAD CITIZENSHIP ON CONSERVATION

Many more people could become purposefully aroused in the cause of conservation if they could appreciate that their positive feelings towards Nature had a genetic basis derived from our evolutionary experience as a species. But there may be a second factor, one at least as important, preventing people who really "know better" from adopting a more active, indeed a more proactive, approach to Nature conservation. The trouble is that we live in a period of "bad citizenship," meaning that it is a time when large masses of ordinary people of good will decide not to engage themselves seriously in matters of fundamental social significance, even if they know that they should. Many people are turned in on themselves, caring mostly for their own interests or those of family and close friends, and for their possessions and "living standards." People may live in a democratic state, the cherished dream of many who do not, but decline to use their freedom to express themselves at the ballot box. Given the long and arduous struggle to arrive at democratic freedoms this is a sign of societal decadence. There seems otherwise no explanation for detachment from the major social issues of the time. It is simply slothfulness in declining to take one's proper part in the responsible activities of a free and just society. This detachment can be a major influence in neglecting to participate in environmental stewardship, in failing to do what one instinctively knows should be done towards the conservation of Nature. Only by having a firm view of one's civic responsibilities can a person with basically good, sound views about the times we live in help to ensure that the world of Nature will survive. In the present morally and ethically slack period it is all too easy to leave actions of public needs to others. If would-

141

be conservationists are not seeing provincial and federal authorities shouldering their duties on conservation problems they should dare to say so. People should understand that even if many of our politicians seem either to be neglectful or cynically opportunistic regarding their responsibilities towards Nature, it is not more than a generation since some lawmakers had the courage and skill to enact laws that, if applied, would ensure a much better deal for many aspects of conservation than we are seeing today. It seems very poor that so few years later many legislators and the government departments they control seem unwilling to act according these existing laws, and why it often seems that many more, who are supposed to have a basic concern for the environment, seem to lack the will or energy to insist on the laws being upheld.

Inaction among government agencies seems to come from a mixture of indifference and reduced departmental staffing caused by reduced funding. Many public servants seem to fear offending business interests in guarding against careless damage to environmental resources. But because whistleblowers on environmental abuses are generally so little regarded, it is fair to say that this is a time of a diminished sense of public responsibility—or bad citizenship. However, since my primary aim would be to awaken a slumbering but powerful goodwill towards Nature I don't think it would be wise to berate people for their torpor. Scolding may be sometimes necessary and useful but may not make friends with those we want to encourage or inspire. On the other hand, it should be part of the understanding of advocates for conservation and the environment that they will need to overcome a custom of social inertia—if not actual indifference or cynicism—that makes the task of the activist more difficult than it ought to be.

CHAPTER 11

THE ETHICS OF CONSERVATION

What might bring about a determination to ensure a new deal in the way humans address the lives and well-being of other living wild creatures of the Earth? Not just because it seems somehow a tolerant or "enlightened" attitude for modern societies, or because scientists and humanists have been persuasive about the many needs to prevent careless extinction of threatened and endangered species. No, the more important point is that we should understand that it is ethically wrong to disregard the extinction of natural life forms, either by directly killing them or by damaging or destroying the ecosystems of which they are members.

Various cultures have held that there is something to revere, or regard as sacred, in living creatures and often also in non-living natural entities such as rocks, water, mountains, sun, moon and stars. Such attitudes may be religious in scope or may be those of anyone who respects Nature's creatures and structures. As for organisms, in recent times it was Albert Schweitzer who famously proposed the attitude of "reverence for life." Schweitzer understood that humans could not survive without causing the death of organisms of other species, but suggested that no creature be destroyed wantonly. Many people find it neither desirable nor possible to approach Schweitzer's doctrine in its purest form. Many people will declare that most humans have never thought they needed to be too respectful of the fate of animals or plants, because they have had to kill or harvest them for food or to safeguard their own lives. As for microbes and other simple life forms, only a few would share the extreme version of Schweitzer's position that even these creatures should be spared unless they are direct threats to human life and welfare.

On the other hand, should we think only of those entities that are capable of consciousness or self-awareness as deserving of special consideration: dogs, but not earthworms? Against that view we have the notion, perhaps originally attributable to Alfred Whitehead, that all entities possess some form of consciousness. The origin of this view emanated from the difficulty of proposing that consciousness "arose" as an emergent property of living creatures as they became more organizationally complex during the course of evolution. That their conscious state then becomes characterized by the complexity of organization of the creatures that possess it. It is certainly simpler to imagine that electrons, in their way, possess consciousness.

Putting aside these difficult questions is it possible to arrive at a truly universal respect for organisms other than humans? What basis can there be for something akin to "rights" of non-human organisms? This must happen to some degree if conservationists are to convince a wide public that there are significant philosophical and ethical reasons for protecting natural creatures and, in particular, the ecosystems in which they have evolved, in which they are "players," and which afford them house, home and sustenance.

Those considerations seem in themselves to be grounds enough for sparing many plants and animals wherever possible. All animal and plant species individually, and as parts of ecosystems to which they belong, should attract deep and thoughtful consideration for their actual or potential "usefulness" to our species. Then there is the fact that all living animals have genetic endowments resembling our own. For example, there is very little difference between the DNAs of chimpanzees, gorillas, orangutans and humans.

Primatologists are already aware of the huge significance of studying primates in a comparative fashion because of what we can learn about the various ways in which evolutionary forces have shaped all of them. If we are really interested in the origins of our own behaviour we need to study that of other primates, other mammals, and birds, reptiles and amphibians.

During their existence, humans have had close associations with a multitude of animal and plant species. Many of these have yielded us food, materials for all kinds of construction and shelter, and innumerable substances of medical and nutritional significance. These facts indicate the great utilitarian advantages of conserving as many species of organisms as possible. If we continue to bio-engineer many food products (hopefully with greater skill and caution as to results than at present), then it is vital that we understand as much as we can of the genetics and biochemistry of the greatest possible variety of plant and animal species.

Despite our deplorable waste and mismanagement of the world's forests, it has been held for many years that by conserving as little as 10 to 12 per cent of original forest we can safeguard the future of representative examples of world forest ecosystems from extinction. However, many forest conservationists are recommending closer to 50 per cent of the natural forests that remain for any worthwhile safeguarding to be effective.

Then there is the problem that depends on a basic ethical attitude towards animals that is often referred to as their "rights," a concept often challenged by belittling intelligence, language and sense of self among animals. Some believe that without these attributes, though animals should be spared pain, hunger and general mistreatment, there is no reason to consider that they possess rights that in any way resemble those of humans.

Yet few who have given the matter thought or study would dispute that (at the least) primates and dogs do display levels of intelligence roughly on par with those attributed to very young children. We do not deny that children have rights which can be legally defended, even if the children cannot speak on their own behalf. Why, then, should not animals be accorded some rights, even if it must be adult humans who will define and defend those rights as they do for children? Go a step further. Many "lower" animals may be unable to effectively demonstrate their own self-awareness. But neither can babies nor many intellectually impaired humans, yet we recognize they have rights. Why then do we arbitrarily decide that lower life forms are undeserving of respect? Back to Schweitzer. He understood.

Most of us who are honest, whether we are devoutly religious or unconvinced sceptics, will agree that, for all our recently-attained scientific knowledge, life, the earth, our planetary system and the universe are still matters of profound mystery. The animal and plant life of ecosystems include the most complex, subtly integrated communities of living creatures we shall ever encounter on this earth. Conservationists must oppose any decisions not to safeguard what remains of them, and consider with extreme care any actions that are supposedly "in the public interest" which may damage or destroy them. These are the creatures of the world other than ourselves. If we resign ourselves to a world of the future that is covered by cities, huge farms of monocultured and genetically-engineered crops, factories and superhighways, we shall have placed the mystery, contemplation and study of all of living Nature forever outside ourselves, out of reach. We shall become vessels empty of all but the feverish glow of our own colossal pride, with nothing left against which to measure the volume of our emptiness. Let us hesitate before we go down that bad road to such a future.

Think for a moment of trees. There are those who can see some case for their conservation but who flinch away from "tree huggers." This seems strange. Faced with some forest giant, perhaps for the first time in their experience that has been formerly confined to city or urban living, does it not seem quite a natural, appropriate and instinctive response to place a hand respectfully on its bark-covered trunk, with the same feeling of wonder, admiration and affection that might lead one to pat the flank of an elephant or to rest one's hand reflectively on a sun-warmed rock? And since a tree is rooted, motionless, harmless, defenceless, why should one not feel inclined to spread arms around its surface? It is certainly with feelings of awe, affection and wonder, as one listens to the soughing of wind in a tree's higher branches, that make some of us at least query why so many do not hesitate to sink their axes or chainsaws into some creature that may have been alive a hundred years before the oldest humans now alive on the planet were conceived.

If, as some insist, humans in times past did often revere trees even when they sometimes cut them to make fire, shelter, boats, utensils, tools and weapons, why has this impulse to respond to trees now become so degraded? There are perhaps two reasons. The first is the invention of the metal axe. When no more than stone tools existed, the cutting of trees was very difficult work. To down a single tree of any great size was likely to have been the slow labour of several hands, as was the shaping of resulting wood into the many forms required. Real thought would have gone into the selection of each tree for felling. Metal axes, and all the tools that have followed them, removed the need for such selective care and rendered the felling and subsequent use of large trees a possible task for a single pair of hands. Wood industries spring up and much wood can be cut for sale or distribution to many others who will use the dead product of the tree without a moment's need to consider its living source. Civilizations are founded on the use by many of dead trees, birds and mammals. Used by crowds, eventually by multitudes, who never see, nor need to think of them alive.

These are the facts and problems of separation between the core of the consciousness of humans as they evolved that so many of us have long ago misplaced and often now reject. The task of those who feel the need to raise again the question of respect for non-human life entails the problem of reminding people of the origins of their species and helping them reawaken sensitivities to Nature that are not dead, only slumbering within. Conservation of Nature will hang in the balance until it has become possible to awaken those feelings in these multitudes.

The thoughtful conservationist will often ponder what it really means to be one. For it is so easy to realize that the pure Nature of so long ago was of a time before technology and that it is technology and the industry that drives it which have spoiled the natural world. This may be correct but it is also too easy. Of course, humans in a state of nature as hunter-gatherers, and before they had devised any tools whatever, could have inflicted no more damage on ecosystems than any other wild animals. They were face to face with Nature. But once the first axe struck its first blow technology existed and our impact was forever observable. Each era of technology built on the shoulders of preceding eras. There were always industrial revolutions. We can neither disown nor repudiate technology. We can only recognize it, understand its impact and safeguard against its more dire effects and prevent them from happening. The problem is, and always has been, to decide whether we will use technology or be ruled by it.

The choice seems so simple …

POSTSCRIPT

Millions are eager to conserve what remains of the living organisms and other features and entities of the natural world. But, even among many of the leaders of thought and action in conservation, increasing numbers feel their cause is being lost for want of time and resources to achieve its goals. Another problem is the lack of the necessary multitude of dedicated volunteers. It will take a veritable army of enthusiasts to help ensure that conservation will not fail.

The current devolution of government conservation agencies means that interested amateurs are increasingly expected to somehow raise the funds to carry out conservation projects. But the funds available are meagre compared to what many of the now devolved government agencies were once able to command, and volunteer conservationists, even when they are enthusiastic and dedicated, frequently lack the necessary experience and insight to make their efforts successful. Their failures will usually occur because they lack ecological experience and training; ecology should always be an essential in the toolkit of conservation. The earlier chapters of this book are a first person narrative of how I became aware of the splendour of the natural world and gradually became an ecologist. Luck and opportunity in my surroundings helped me, though at the beginning I lacked the sort of wide exposure to ecological science that is now a normal part of biological education.

I believe that a narrative recounting of how an ecological viewpoint may be acquired can provide substantial help in clarifying the real underpinnings of conservation problems—an understanding that is vital for anyone striving to become an active conservationist. The narrative begins in my early childhood interest in Nature and provides examples of increasing experience and understanding that extends to the present. I think this will get to the point of ecological and conservationist understanding among amateurs better than by their diving directly into formal textbook treatments of these subjects, which may require additional scientific knowledge likely to be missing among so many.

In the book's second, non-narrative, part there is a brief history of humans' relationship with Nature, based on a story of forests and the desperate need to avoid their total devastation. Forests should be regarded as hugely important parts of the Nature that is, in ecological terms, "our one true home," to which we are related through evolution, history, economics and spirit. The challenge is to understand what happened as the human species moved away from forests and from rivers, mountains and wildlife—and what it will take to refocus our concern for these things at both local and global scales. Accordingly, the book claims that the most basic and vital purpose of conservationists must be to address the usually overlooked evidence that humans have a deep, instinctual (but commonly mislaid) love of Nature, and that this love can be reawakened. If it is recognized that this is possible, the conservation movement can become revitalized in the form of a vast, very evident and very powerful global concern for the planet's welfare, so pervasively influential that neither business nor politics will be able to resist it.

It is particularly important to discourage the notion that in this day and age we must concentrate on industry and wealth rather than mere conservation of Nature. In fact, it is the condition of the natural world that will ultimately always determine the overall quality of life in the human global population.

Conservation applies to our societies as well as to the other entities of the world. Soon we will have to examine the ways in which societies have interacted with each other in utilizing the planet's plummeting stores of raw materials in our long and essentially unending struggle for wealth. History clearly shows that many human interactions have been grossly unfair and have led to, or resulted from, monstrous misjudgements and misbehaviours between societies. I have treated this subject in a fashion that moves away from traditional economics and assumes a more ecological standpoint in confronting the enduring pressures between societies. It suggests, regardless of ideologically-coloured views of the present world, that the future will require a much greater organization in the ways we live on and treat the Earth.

The book also contains short essays on the meaning of land ownership, "good and bad" citizenship, the ethics of conservation and reconsidering the meaning of the struggle for conservation.

ACKNOWLEDGEMENTS

In preparing this book I have drawn on various source materials, some from my own previous publications, others works of other authors. In the text I have endeavoured to provide sufficient detail of books referred to, to allow readers interested in seeking out these works, to readily find them via the Internet. In the case of the Canaan River and Lake Washademoak studies, there are various publications that may interest local environment groups. Items may be found at http://www.nbwatersheds.ca/cwwa and associated links. Access to any items not on the website may be obtained by contacting Canaan-Washademoak Watershed Association, 25 Colonial Heights, Fredericton, New Brunswick, Canada E3B 5M2.

I am extremely grateful to Bohdan Harasymiw who provided crucial suggestions early in the development of the book. Later, his help editing the final version was invaluable.

Thanks to Lisa Jeffrey and Robena Weatherley for critical editorial assistance.

Thanks also to Ian Varty for technical assistance and to Julian Varty for careful reading of the manuscript.